The Boson

Edited by Paul F. Kisak

Contents

Chapter 1

Boson

For other uses, see Boson (disambiguation).

In quantum mechanics, a **boson** (/'boʊsɒn/,[1] /'boʊzɒn/[2]) is a particle that follows Bose–Einstein statistics. Bosons make up one of the two classes of particles, the other being fermions.[3] The name boson was coined by Paul Dirac[4] to commemorate the contribution of the Indian physicist Satyendra Nath Bose[5][6] in developing, with Einstein, Bose–Einstein statistics—which theorizes the characteristics of elementary particles.[7] Examples of bosons include fundamental particles such as photons, gluons, and W and Z bosons (the four force-carrying gauge bosons of the Standard Model), the Higgs boson, and the still-theoretical graviton of quantum gravity; composite particles (e.g. mesons and stable nuclei of even mass number such as deuterium (with one proton and one neutron, mass number = 2), helium-4, or lead-208[Note 1]); and some quasiparticles (e.g. Cooper pairs, plasmons, and phonons).[8]:130

An important characteristic of bosons is that their statistics do not restrict the number of them that occupy the same quantum state. This property is exemplified by helium-4 when it is cooled to become a superfluid.[9] Unlike bosons, two identical fermions cannot occupy the same quantum space. Whereas the elementary particles that make up matter (i.e. leptons and quarks) are fermions, the elementary bosons are force carriers that function as the 'glue' holding matter together.[10] This property holds for all particles with integer spin (s = 0, 1, 2 etc.) as a consequence of the spin–statistics theorem. When a gas of Bose particles is cooled down to absolute zero then the kinetic energy of the particles decreases to a negligible amount and they condense into a lowest energy level state. This state is called Bose Einstein condensation. It is believed that this phenomenon is the secret behind superfluidity of liquids.

1.1 Types

Bosons may be either elementary, like photons, or composite, like mesons.

While most bosons are composite particles, in the Standard Model there are five bosons which are elementary:

- the four gauge bosons ($\gamma \cdot g \cdot Z \cdot W\pm$)

- the only scalar boson (the Higgs boson (H0))

Additionally, the graviton (G) is a hypothetical elementary particle not incorporated in the Standard Model. If it exists, a graviton must be a boson, and could conceivably be a gauge boson.

Composite bosons are important in superfluidity and other applications of Bose–Einstein condensates. When a gas of Bose particles is cooled to absolute zero its kinetic energy decreases up to a negligible amount then the particles would condense into the lowest energy state. This phenomenon is known as Bose-Einstein condensation and it is believed that this phenomenon is the secret behind superfluidity of liquids.

Satyendra Nath Bose

1.2 Properties

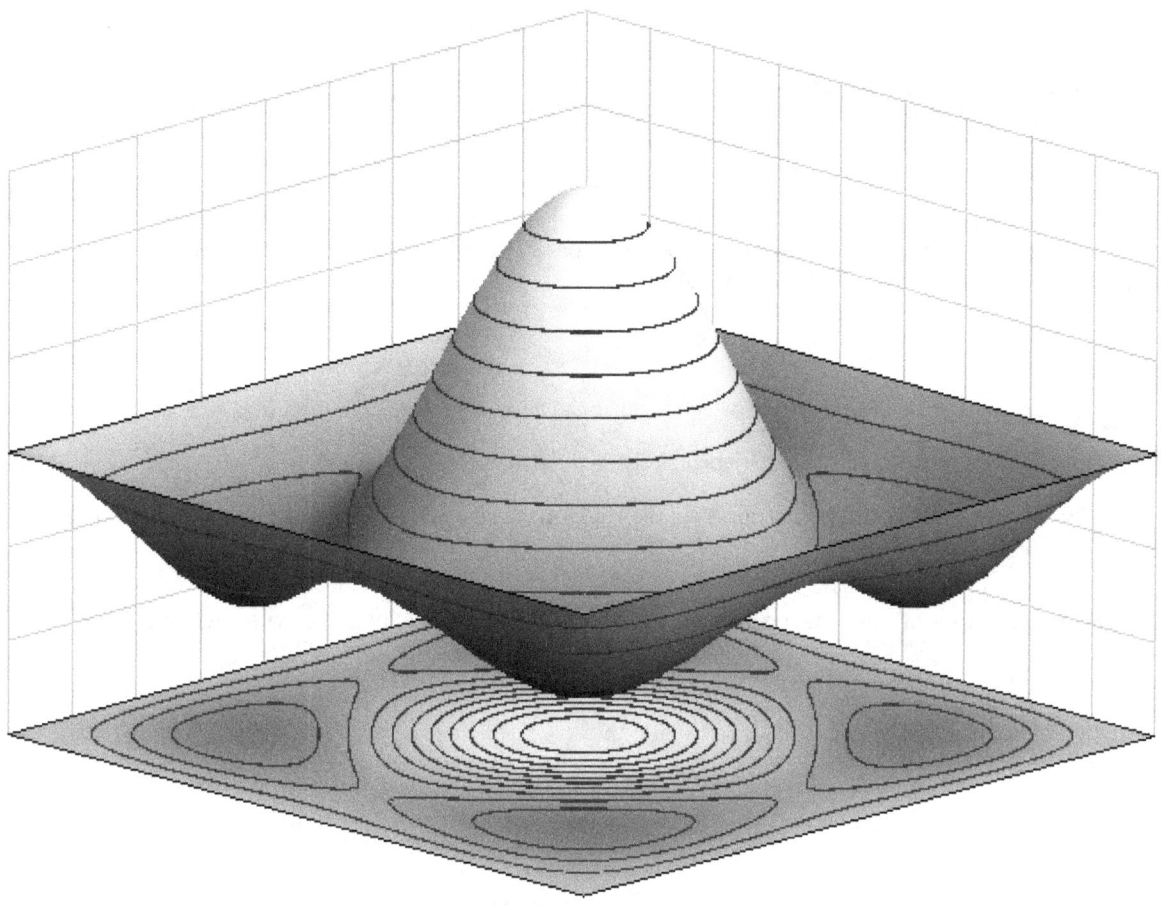

Symmetric wavefunction for a (bosonic) 2-particle state in an infinite square well potential.

Bosons differ from fermions, which obey Fermi–Dirac statistics. Two or more identical fermions cannot occupy the same quantum state (see Pauli exclusion principle).

Since bosons with the same energy can occupy the same place in space, bosons are often force carrier particles. Fermions are usually associated with matter (although in quantum physics the distinction between the two concepts is not clear cut).

Bosons are particles which obey Bose–Einstein statistics: when one swaps two bosons (of the same species), the wavefunction of the system is unchanged.[11] Fermions, on the other hand, obey Fermi–Dirac statistics and the Pauli exclusion principle: two fermions cannot occupy the same quantum state, resulting in a "rigidity" or "stiffness" of matter which includes fermions. Thus fermions are sometimes said to be the constituents of matter, while bosons are said to be the particles that transmit interactions (force carriers), or the constituents of radiation. The quantum fields of bosons are bosonic fields, obeying canonical commutation relations.

The properties of lasers and masers, superfluid helium-4 and Bose–Einstein condensates are all consequences of statistics of bosons. Another result is that the spectrum of a photon gas in thermal equilibrium is a Planck spectrum, one example of which is black-body radiation; another is the thermal radiation of the opaque early Universe seen today as microwave background radiation. Interactions between elementary particles are called fundamental interactions. The fundamental interactions of virtual bosons with real particles result in all forces we know.

All known elementary and composite particles are bosons or fermions, depending on their spin: particles with half-integer spin are fermions; particles with integer spin are bosons. In the framework of nonrelativistic quantum mechanics, this is a purely empirical observation. However, in relativistic quantum field theory, the spin–statistics theorem shows that half-integer spin particles cannot be bosons and integer spin particles cannot be fermions.[12]

In large systems, the difference between bosonic and fermionic statistics is only apparent at large densities—when their wave functions overlap. At low densities, both types of statistics are well approximated by Maxwell–Boltzmann statistics, which is described by classical mechanics.

1.3   Elementary bosons

See also: List of particles: Bosons

All observed elementary particles are either fermions or bosons. The observed elementary bosons are all gauge bosons: photons, W and Z bosons, gluons, and the Higgs boson.

- Photons are the force carriers of the electromagnetic field.

- W and Z bosons are the force carriers which mediate the weak force.

- Gluons are the fundamental force carriers underlying the strong force.

- Higgs Bosons give W and Z bosons mass via the Higgs mechanism. Their existence was confirmed by CERN on 14 March 2013.

Finally, many approaches to quantum gravity postulate a force carrier for gravity, the graviton, which is a boson of spin plus or minus two.

1.4   Composite bosons

See also: List of particles: Composite particles

Composite particles (such as hadrons, nuclei, and atoms) can be bosons or fermions depending on their constituents. More precisely, because of the relation between spin and statistics, a particle containing an even number of fermions is a boson, since it has integer spin.

Examples include the following:

- Any meson, since mesons contain one quark and one antiquark.

- The nucleus of a carbon-12 atom, which contains 6 protons and 6 neutrons.

- The helium-4 atom, consisting of 2 protons, 2 neutrons and 2 electrons.

The number of bosons within a composite particle made up of simple particles bound with a potential has no effect on whether it is a boson or a fermion.

1.5   To which states can bosons crowd?

Bose–Einstein statistics encourages identical bosons to crowd into one quantum state, but not any state is necessarily convenient for it. Aside of statistics, bosons can interact – for example, helium-4 atoms are repulsed by intermolecular force on a very close approach, and if one hypothesizes their condensation in a spatially-localized state, then gains from the statistics cannot overcome a prohibitive force potential. A spatially-delocalized state (i.e. with low $|\psi(x)|$) is preferable: if the number density of the condensate is about the same as in ordinary liquid or solid state, then the repulsive potential for the N-particle condensate in such state can be not higher than for a liquid or a crystalline lattice of the same N particles described without quantum statistics. Thus, Bose–Einstein statistics for a material particle is not a mechanism to bypass physical restrictions on the density of the corresponding substance, and superfluid liquid helium has the density comparable to the density of ordinary liquid matter. Spatially-delocalized states also permit for a low momentum according to uncertainty principle, hence for low kinetic energy; that's why superfluidity and superconductivity are usually observed in low temperatures.

Photons do not interact with themselves and hence do not experience this difference in states where to crowd (see squeezed coherent state).

1.6 See also

- Anyon
- Bose gas
- Identical particles
- Parastatistics
- Fermion

1.7 Notes

[1] Even-mass-number nuclides, which comprise 152/255 = ~ 60% of all stable nuclides, are bosons, i.e. they have integer spin. Almost all (148 of the 152) are even-proton, even-neutron (EE) nuclides, which necessarily have spin 0 because of pairing. The remainder of the stable bosonic nuclides are 5 odd-proton, odd-neutron stable nuclides (see even and odd atomic nuclei#Odd proton, odd neutron); these odd–odd bosons are: 2
1H, 6
3Li,10
5B, 14
7N and 180m
73Ta). All have nonzero integer spin.

1.8 References

[1] Wells, John C. (1990). *Longman pronunciation dictionary*. Harlow, England: Longman. ISBN 0582053838. entry "Boson"

[2] "boson". *Collins Dictionary*.

[3] Carroll, Sean (2007) *Dark Matter, Dark Energy: The Dark Side of the Universe*, Guidebook Part 2 p. 43, The Teaching Company, ISBN 1598033506 "...boson: A force-carrying particle, as opposed to a matter particle (fermion). Bosons can be piled on top of each other without limit. Examples include photons, gluons, gravitons, weak bosons, and the Higgs boson. The spin of a boson is always an integer, such as 0, 1, 2, and so on..."

[4] Notes on Dirac's lecture *Developments in Atomic Theory* at Le Palais de la Découverte, 6 December 1945, UKNATARCHI Dirac Papers BW83/2/257889. See note 64 to p. 331 in "The Strangest Man" by Graham Farmelo

[5] Daigle, Katy (10 July 2012). "India: Enough about Higgs, let's discuss the boson". *AP News*. Retrieved 10 July 2012.

[6] Bal, Hartosh Singh (19 September 2012). "The Bose in the Boson". *New York Times blog*. Retrieved 21 September 2012.

[7] "Higgs boson: The poetry of subatomic particles". *BBC News*. 4 July 2012. Retrieved 6 July 2012.

[8] Charles P. Poole, Jr. (11 March 2004). *Encyclopedic Dictionary of Condensed Matter Physics*. Academic Press. ISBN 978-0-08-054523-3.

[9] "boson". *Merriam-Webster Online Dictionary*. Retrieved 21 March 2010.

[10] Carroll, Sean. "Explain it in 60 seconds: Bosons". *Symmetry Magazine*. Fermilab/SLAC. Retrieved 15 February 2013.

[11] Srednicki, Mark (2007). *Quantum Field Theory*, Cambridge University Press, pp. 28–29, ISBN 978-0-521-86449-7.

[12] Sakurai, J.J. (1994). *Modern Quantum Mechanics* (Revised Edition), p. 362. Addison-Wesley, ISBN 0-201-53929-2.

Chapter 2

Gauge boson

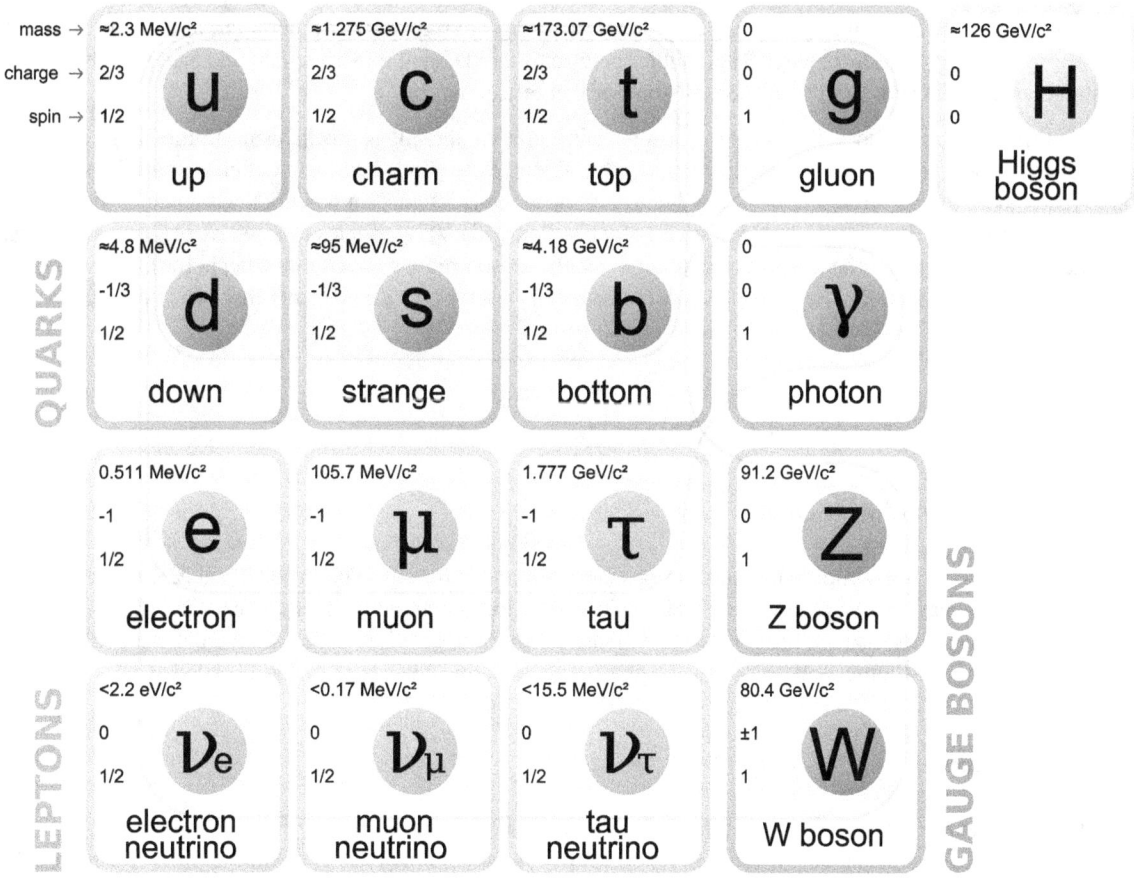

The Standard Model of elementary particles, with the gauge bosons in the fourth column in red

In particle physics, a **gauge boson** is a force carrier, a bosonic particle that carries any of the fundamental interactions of nature.[1][2] Elementary particles, whose interactions are described by a gauge theory, interact with each other by the exchange of gauge bosons—usually as virtual particles.

2.1   Gauge bosons in the Standard Model

The Standard Model of particle physics recognizes four kinds of gauge bosons: photons, which carry the electromagnetic interaction; W and Z bosons, which carry the weak interaction; and gluons, which carry the strong interaction.[3]

Isolated gluons do not occur at low energies because they are color-charged, and subject to color confinement.

2.1.1   Multiplicity of gauge bosons

In a quantized gauge theory, gauge bosons are quanta of the gauge fields. Consequently, there are as many gauge bosons as there are generators of the gauge field. In quantum electrodynamics, the gauge group is $U(1)$; in this simple case, there is only one gauge boson. In quantum chromodynamics, the more complicated group $SU(3)$ has eight generators, corresponding to the eight gluons. The three W and Z bosons correspond (roughly) to the three generators of $SU(2)$ in GWS theory.

2.1.2   Massive gauge bosons

For technical reasons involving gauge invariance, gauge bosons are described mathematically by field equations for massless particles. Therefore, at a naïve theoretical level all gauge bosons are required to be massless, and the forces that they describe are required to be long-ranged. The conflict between this idea and experimental evidence that the weak and strong interactions have a very short range requires further theoretical insight.

According to the Standard Model, the W and Z bosons gain mass via the Higgs mechanism. In the Higgs mechanism, the four gauge bosons (of $SU(2) \times U(1)$ symmetry) of the unified electroweak interaction couple to a Higgs field. This field undergoes spontaneous symmetry breaking due to the shape of its interaction potential. As a result, the universe is permeated by a nonzero Higgs vacuum expectation value (VEV). This VEV couples to three of the electroweak gauge bosons (the Ws and Z), giving them mass; the remaining gauge boson remains massless (the photon). This theory also predicts the existence of a scalar Higgs boson, which has been observed in experiments that were reported on 4 July 2012.[4]

2.2   Beyond the Standard Model

2.2.1   Grand unification theories

A grand unified theory predicts additional gauge bosons named X and Y bosons. The hypothetical X and Y bosons direct interactions between quarks and leptons, hence violating conservation of baryon number and causing proton decay. Such bosons would be even more massive than W and Z bosons due to symmetry breaking. Analysis of data collected from such sources as the Super-Kamiokande neutrino detector has yielded no evidence of X and Y bosons.

2.2.2   Gravitons

The fourth fundamental interaction, gravity, may also be carried by a boson, called the graviton. In the absence of experimental evidence and a mathematically coherent theory of quantum gravity, it is unknown whether this would be a gauge boson or not. The role of gauge invariance in general relativity is played by a similar symmetry: diffeomorphism invariance.

2.2.3   W' and Z' bosons

Main article: W' and Z' bosons

W' and Z' bosons refer to hypothetical new gauge bosons (named in analogy with the Standard Model W and Z bosons).

2.3 See also

- 1964 PRL symmetry breaking papers
- Boson
- Glueball
- Quantum chromodynamics
- Quantum electrodynamics

2.4 References

[1] Gribbin, John (2000). *Q is for Quantum – An Encyclopedia of Particle Physics*. Simon & Schuster. ISBN 0-684-85578-X.

[2] Clark, John, E.O. (2004). *The Essential Dictionary of Science*. Barnes & Noble. ISBN 0-7607-4616-8.

[3] Veltman, Martinus (2003). *Facts and Mysteries in Elementary Particle Physics*. World Scientific. ISBN 981-238-149-X.

[4] "CERN experiments observe particle consistent with long-sought Higgs boson". CERN. Retrieved 4 July 2012.

2.5 External links

- Explanation of gauge boson and gauge fields by Christopher T. Hill

Chapter 3

Photon

This article is about the elementary particle of light. For other uses, see Photon (disambiguation).

A **photon** is an elementary particle, the quantum of light and all other forms of electromagnetic radiation. It is the force carrier for the electromagnetic force, even when static via virtual photons. The effects of this force are easily observable at the microscopic and at the macroscopic level, because the photon has zero rest mass; this allows long distance interactions. Like all elementary particles, photons are currently best explained by quantum mechanics and exhibit wave–particle duality, exhibiting properties of waves and of particles. For example, a single photon may be refracted by a lens or exhibit wave interference with itself, but also act as a particle giving a definite result when its position is measured. Waves and quanta, being two observable aspects of a single phenomenon cannot have their true nature described in terms of any mechanical model. [2] A representation of this dual property of light, which assumes certain points on the wave front to be the seat of the energy is also impossible. Thus, the quanta in a light wave cannot be spatially localized. Some defined physical parameters of a photon are listed.

The modern photon concept was developed gradually by Albert Einstein in the first years of the 20th century to explain experimental observations that did not fit the classical wave model of light. In particular, the photon model accounted for the frequency dependence of light's energy, and explained the ability of matter and radiation to be in thermal equilibrium. It also accounted for anomalous observations, including the properties of black-body radiation, that other physicists, most notably Max Planck, had sought to explain using *semiclassical models*, in which light is still described by Maxwell's equations, but the material objects that emit and absorb light do so in amounts of energy that are *quantized* (i.e., they change energy only by certain particular discrete amounts and cannot change energy in any arbitrary way). Although these semiclassical models contributed to the development of quantum mechanics, many further experiments[3][4] starting with Compton scattering of single photons by electrons, first observed in 1923, validated Einstein's hypothesis that *light itself* is quantized. In 1926 the optical physicist Frithiof Wolfers and the chemist Gilbert N. Lewis coined the name *photon* for these particles, and after 1927, when Arthur H. Compton won the Nobel Prize for his scattering studies, most scientists accepted the validity that quanta of light have an independent existence, and the term *photon* for light quanta was accepted.

In the Standard Model of particle physics, photons and other elementary particles are described as a necessary consequence of physical laws having a certain symmetry at every point in spacetime. The intrinsic properties of particles, such as charge, mass and spin, are determined by the properties of this gauge symmetry. The photon concept has led to momentous advances in experimental and theoretical physics, such as lasers, Bose–Einstein condensation, quantum field theory, and the probabilistic interpretation of quantum mechanics. It has been applied to photochemistry, high-resolution microscopy, and measurements of molecular distances. Recently, photons have been studied as elements of quantum computers and for applications in optical imaging and optical communication such as quantum cryptography.

3.1 Nomenclature

In 1900, the German physicist Max Planck was working on black-body radiation and suggested that the energy in electromagnetic waves could only be released in "packets" of energy. In his 1901 article [5] in Annalen der Physik he called these packets "energy elements". The word *quanta* (singular *quantum*) was used even before 1900 to mean particles or amounts of different quantities, including electricity. Later, in 1905, Albert Einstein went further by suggesting that electromagnetic waves could only exist in these discrete wave-packets.[6] He called such a wave-packet *the light quantum* (German: *das Lichtquant*).[Note 1] The name *photon* derives from the Greek word for light, φῶς (transliterated *phôs*). Arthur Compton used *photon* in 1928, referring to Gilbert N. Lewis.[7] The same name was used earlier, by the American physicist and psychologist Leonard T. Troland, who coined the word in 1916, in 1921 by the Irish physicist John Joly, in 1924 by the French physiologist René Wurmser (1890-1993) and in 1926 by the French physicist Frithiof Wolfers (1891-1971).[8] The name was suggested initially as a unit related to the illumination of the eye and the resulting sensation of light and was used later on in a physiological context. Although Wolfers's and Lewis's theories were never accepted, as they were contradicted by many experiments, the new name was adopted very soon by most physicists after Compton used it.[8][Note 2]

In physics, a photon is usually denoted by the symbol γ (the Greek letter gamma). This symbol for the photon probably derives from gamma rays, which were discovered in 1900 by Paul Villard,[9][10] named by Ernest Rutherford in 1903, and shown to be a form of electromagnetic radiation in 1914 by Rutherford and Edward Andrade.[11] In chemistry and optical engineering, photons are usually symbolized by $h\nu$, the energy of a photon, where h is Planck's constant and the Greek letter ν (nu) is the photon's frequency. Much less commonly, the photon can be symbolized by hf, where its frequency is denoted by f.

3.2 Physical properties

See also: Special relativity and Photonic molecule

A photon is massless,[Note 3] has no electric charge,[12] and is stable. A photon has two possible polarization states. In the momentum representation, which is preferred in quantum field theory, a photon is described by its wave vector, which determines its wavelength λ and its direction of propagation. A photon's wave vector may not be zero and can be represented either as a spatial 3-vector or as a (relativistic) four-vector; in the latter case it belongs to the light cone (pictured). Different signs of the four-vector denote different circular polarizations, but in the 3-vector representation one should account for the polarization state separately; it actually is a spin quantum number. In both cases the space of possible wave vectors is three-dimensional.

The photon is the gauge boson for electromagnetism,[13]:29-30 and therefore all other quantum numbers of the photon (such as lepton number, baryon number, and flavour quantum numbers) are zero.[14] Also, the photon does not obey the Pauli exclusion principle.[15]:1221

Photons are emitted in many natural processes. For example, when a charge is accelerated it emits synchrotron radiation. During a molecular, atomic or nuclear transition to a lower energy level, photons of various energy will be emitted, from radio waves to gamma rays. A photon can also be emitted when a particle and its corresponding antiparticle are annihilated (for example, electron–positron annihilation).[15]:572, 1114, 1172

In empty space, the photon moves at c (the speed of light) and its energy and momentum are related by $E = pc$, where p is the magnitude of the momentum vector **p**. This derives from the following relativistic relation, with $m = 0$:[16]

$$E^2 = p^2 c^2 + m^2 c^4.$$

The energy and momentum of a photon depend only on its frequency (ν) or inversely, its wavelength (λ):

$$E = \hbar\omega = h\nu = \frac{hc}{\lambda}$$

$$\boldsymbol{p} = \hbar\boldsymbol{k},$$

where k is the wave vector (where the wave number $k = |k| = 2\pi/\lambda$), $\omega = 2\pi\nu$ is the angular frequency, and $\hbar = h/2\pi$ is the reduced Planck constant.[17]

Since p points in the direction of the photon's propagation, the magnitude of the momentum is

$$p = \hbar k = \frac{h\nu}{c} = \frac{h}{\lambda}.$$

The photon also carries spin angular momentum that does not depend on its frequency.[18] The magnitude of its spin is $\sqrt{2}\hbar$ and the component measured along its direction of motion, its helicity, must be $\pm\hbar$. These two possible helicities, called right-handed and left-handed, correspond to the two possible circular polarization states of the photon.[19]

To illustrate the significance of these formulae, the annihilation of a particle with its antiparticle in free space must result in the creation of at least *two* photons for the following reason. In the center of momentum frame, the colliding antiparticles have no net momentum, whereas a single photon always has momentum (since, as we have seen, it is determined by the photon's frequency or wavelength, which cannot be zero). Hence, conservation of momentum (or equivalently, translational invariance) requires that at least two photons are created, with zero net momentum. (However, it is possible if the system interacts with another particle or field for annihilation to produce one photon, as when a positron annihilates with a bound atomic electron, it is possible for only one photon to be emitted, as the nuclear Coulomb field breaks translational symmetry.)[20]:64-65 The energy of the two photons, or, equivalently, their frequency, may be determined from conservation of four-momentum. Seen another way, the photon can be considered as its own antiparticle. The reverse process, pair production, is the dominant mechanism by which high-energy photons such as gamma rays lose energy while passing through matter.[21] That process is the reverse of "annihilation to one photon" allowed in the electric field of an atomic nucleus.

The classical formulae for the energy and momentum of electromagnetic radiation can be re-expressed in terms of photon events. For example, the pressure of electromagnetic radiation on an object derives from the transfer of photon momentum per unit time and unit area to that object, since pressure is force per unit area and force is the change in momentum per unit time.[22]

3.2.1   Experimental checks on photon mass

Current commonly accepted physical theories imply or assume the photon to be strictly massless. If the photon is not a strictly massless particle, it would not move at the exact speed of light in vacuum, c. Its speed would be lower and depend on its frequency. Relativity would be unaffected by this; the so-called speed of light, c, would then not be the actual speed at which light moves, but a constant of nature which is the maximum speed that any object could theoretically attain in space-time.[23] Thus, it would still be the speed of space-time ripples (gravitational waves and gravitons), but it would not be the speed of photons.

If a photon did have non-zero mass, there would be other effects as well. Coulomb's law would be modified and the electromagnetic field would have an extra physical degree of freedom. These effects yield more sensitive experimental probes of the photon mass than the frequency dependence of the speed of light. If Coulomb's law is not exactly valid, then that would cause the presence of an electric field inside a hollow conductor when it is subjected to an external electric field. This thus allows one to test Coulomb's law to very high precision.[24] A null result of such an experiment has set a limit of $m \lesssim 10^{-14}$ eV/c^2.[25]

Sharper upper limits have been obtained in experiments designed to detect effects caused by the galactic vector potential. Although the galactic vector potential is very large because the galactic magnetic field exists on very long length scales, only the magnetic field is observable if the photon is massless. In case of a massive photon, the mass term $\frac{1}{2}m^2 A_\mu A^\mu$ would affect the galactic plasma. The fact that no such effects are seen implies an upper bound on the photon mass of $m < 3\times10^{-27}$ eV/c^2.[26] The galactic vector potential can also be probed directly by measuring the torque exerted on a magnetized ring.[27] Such methods were used to obtain the sharper upper limit of 10^{-18}eV/c^2 (the equivalent of 1.07×10^{-27} atomic mass units) given by the Particle Data Group.[28]

These sharp limits from the non-observation of the effects caused by the galactic vector potential have been shown to be model dependent.[29] If the photon mass is generated via the Higgs mechanism then the upper limit of $m\lesssim10^{-14}$ eV/c^2 from the test of Coulomb's law is valid.

Photons inside superconductors do develop a nonzero effective rest mass; as a result, electromagnetic forces become short-range inside superconductors.[30]

See also: Supernova/Acceleration Probe

3.3 Historical development

Main article: Light

In most theories up to the eighteenth century, light was pictured as being made up of particles. Since particle models cannot easily account for the refraction, diffraction and birefringence of light, wave theories of light were proposed by René Descartes (1637),[31] Robert Hooke (1665),[32] and Christiaan Huygens (1678);[33] however, particle models remained dominant, chiefly due to the influence of Isaac Newton.[34] In the early nineteenth century, Thomas Young and August Fresnel clearly demonstrated the interference and diffraction of light and by 1850 wave models were generally accepted.[35] In 1865, James Clerk Maxwell's prediction[36] that light was an electromagnetic wave—which was confirmed experimentally in 1888 by Heinrich Hertz's detection of radio waves[37]—seemed to be the final blow to particle models of light.

The Maxwell wave theory, however, does not account for *all* properties of light. The Maxwell theory predicts that the energy of a light wave depends only on its intensity, not on its frequency; nevertheless, several independent types of experiments show that the energy imparted by light to atoms depends only on the light's frequency, not on its intensity. For example, some chemical reactions are provoked only by light of frequency higher than a certain threshold; light of frequency lower than the threshold, no matter how intense, does not initiate the reaction. Similarly, electrons can be ejected from a metal plate by shining light of sufficiently high frequency on it (the photoelectric effect); the energy of the ejected electron is related only to the light's frequency, not to its intensity.[38][Note 4]

At the same time, investigations of blackbody radiation carried out over four decades (1860–1900) by various researchers[culminated inMax Planck'shypothesis[5][40]that the energy ofanysystem that absorbs or emits electromagnetic radiation of frequency$v$is an integer multiple of an energy quantum $E = hv$. As shown by Albert Einstein,[6][41] some form of energy quantization *must* be assumed to account for the thermal equilibrium observed between matter and electromagnetic radiation; for this explanation of the photoelectric effect, Einstein received the 1921 Nobel Prize in physics.[42]

Since the Maxwell theory of light allows for all possible energies of electromagnetic radiation, most physicists assumed initially that the energy quantization resulted from some unknown constraint on the matter that absorbs or emits the radiation. In 1905, Einstein was the first to propose that energy quantization was a property of electromagnetic radiation itself.[6] Although he accepted the validity of Maxwell's theory, Einstein pointed out that many anomalous experiments could be explained if the *energy* of a Maxwellian light wave were localized into point-like quanta that move independently of one another, even if the wave itself is spread continuously over space.[6] In 1909[41] and 1916,[43] Einstein showed that, if Planck's law of black-body radiation is accepted, the energy quanta must also carry momentum $p = h/\lambda$, making them full-fledged particles. This photon momentum was observed experimentally[44] by Arthur Compton, for which he received the Nobel Prize in 1927. The pivotal question was then: how to unify Maxwell's wave theory of light with its experimentally observed particle nature? The answer to this question occupied Albert Einstein for the rest of his life,[45] and was solved in quantum electrodynamics and its successor, the Standard Model (see Second quantization and The photon as a gauge boson, below).

3.4 Einstein's light quantum

Unlike Planck, Einstein entertained the possibility that there might be actual physical quanta of light—what we now call photons. He noticed that a light quantum with energy proportional to its frequency would explain a number of troubling puzzles and paradoxes, including an unpublished law by Stokes, the ultraviolet catastrophe, and of course the photoelectric effect. Stokes's law said simply that the frequency of fluorescent light cannot be greater than the frequency of the light (usually ultraviolet) inducing it. Einstein eliminated the ultraviolet catastrophe by imagining a gas of photons behaving like a gas of electrons that he had previously considered. He was advised by a colleague to be careful how he wrote up this paper, in order to not challenge Planck too directly, as he was a powerful figure, and indeed the warning was justified, as Planck never forgave him for writing it.[46]

3.5   Early objections

Einstein's 1905 predictions were verified experimentally in several ways in the first two decades of the 20th century, as recounted in Robert Millikan's Nobel lecture.[47] However, before Compton's experiment[44] showing that photons carried momentum proportional to their wave number (or frequency) (1922), most physicists were reluctant to believe that electromagnetic radiation itself might be particulate. (See, for example, the Nobel lectures of Wien,[39] Planck[40] and Millikan.[47]) Instead, there was a widespread belief that energy quantization resulted from some unknown constraint on the matter that absorbs or emits radiation. Attitudes changed over time. In part, the change can be traced to experiments such as Compton scattering, where it was much more difficult not to ascribe quantization to light itself to explain the observed results.[48]

Even after Compton's experiment, Niels Bohr, Hendrik Kramers and John Slater made one last attempt to preserve the Maxwellian continuous electromagnetic field model of light, the so-called BKS model.[49] To account for the data then available, two drastic hypotheses had to be made:

1. **Energy and momentum are conserved only on the average in interactions between matter and radiation, not in elementary processes such as absorption and emission.** This allows one to reconcile the discontinuously changing energy of the atom (jump between energy states) with the continuous release of energy into radiation.

2. **Causality is abandoned**. For example, spontaneous emissions are merely emissions induced by a "virtual" electromagnetic field.

However, refined Compton experiments showed that energy–momentum is conserved extraordinarily well in elementary processes; and also that the jolting of the electron and the generation of a new photon in Compton scattering obey causality to within 10 ps. Accordingly, Bohr and his co-workers gave their model "as honorable a funeral as possible".[45] Nevertheless, the failures of the BKS model inspired Werner Heisenberg in his development of matrix mechanics.[50]

A few physicists persisted[51] in developing semiclassical models in which electromagnetic radiation is not quantized, but matter appears to obey the laws of quantum mechanics. Although the evidence for photons from chemical and physical experiments was overwhelming by the 1970s, this evidence could not be considered as *absolutely* definitive; since it relied on the interaction of light with matter, a sufficiently complicated theory of matter could in principle account for the evidence. Nevertheless, *all* semiclassical theories were refuted definitively in the 1970s and 1980s by photon-correlation experiments.[Note 5] Hence, Einstein's hypothesis that quantization is a property of light itself is considered to be proven.

3.6   Wave–particle duality and uncertainty principles

See also: Wave–particle duality, Squeezed coherent state, Uncertainty principle and De Broglie–Bohm theory
 Photons, like all quantum objects, exhibit wave-like and particle-like properties. Their dual wave–particle nature can be difficult to visualize. The photon displays clearly wave-like phenomena such as diffraction and interference on the length scale of its wavelength. For example, a single photon passing through a double-slit experiment lands on the screen exhibiting interference phenomena but only if no measure was made on the actual slit being run across. To account for the particle interpretation that phenomenon is called probability distribution but behaves according to Maxwell's equations.[52] However, experiments confirm that the photon is *not* a short pulse of electromagnetic radiation; it does not spread out as it propagates, nor does it divide when it encounters a beam splitter.[53] Rather, the photon seems to be a point-like particle since it is absorbed or emitted *as a whole* by arbitrarily small systems, systems much smaller than its wavelength, such as an atomic nucleus ($\approx 10^{-15}$ m across) or even the point-like electron. Nevertheless, the photon is *not* a point-like particle whose trajectory is shaped probabilistically by the electromagnetic field, as conceived by Einstein and others; that hypothesis was also refuted by the photon-correlation experiments cited above. According to our present understanding, the electromagnetic field itself is produced by photons, which in turn result from a local gauge symmetry and the laws of quantum field theory (see the Second quantization and Gauge boson sections below).

A key element of quantum mechanics is Heisenberg's uncertainty principle, which forbids the simultaneous measurement of the position and momentum of a particle along the same direction. Remarkably, the uncertainty principle for charged, material particles *requires* the quantization of light into photons, and even the frequency dependence of the photon's energy and momentum. An elegant illustration is Heisenberg's thought experiment for locating an electron with an ideal microscope.[54] The position of the electron can be determined to within the resolving power of the microscope, which is given by a formula from classical optics

$$\Delta x \sim \frac{\lambda}{\sin \theta}$$

where θ is the aperture angle of the microscope. Thus, the position uncertainty Δx can be made arbitrarily small by reducing the wavelength λ. The momentum of the electron is uncertain, since it received a "kick" Δp from the light scattering from it into the microscope. If light were *not* quantized into photons, the uncertainty Δp could be made arbitrarily small by reducing the light's intensity. In that case, since the wavelength and intensity of light can be varied independently, one could simultaneously determine the position and momentum to arbitrarily high accuracy, violating the uncertainty principle. By contrast, Einstein's formula for photon momentum preserves the uncertainty principle; since the photon is scattered anywhere within the aperture, the uncertainty of momentum transferred equals

$$\Delta p \sim p_{\text{photon}} \sin \theta = \frac{h}{\lambda} \sin \theta$$

giving the product $\Delta x \Delta p \sim h$, which is Heisenberg's uncertainty principle. Thus, the entire world is quantized; both matter and fields must obey a consistent set of quantum laws, if either one is to be quantized.[55]

The analogous uncertainty principle for photons forbids the simultaneous measurement of the number n of photons (see Fock state and the Second quantization section below) in an electromagnetic wave and the phase ϕ of that wave

$$\Delta n \Delta \phi > 1$$

See coherent state and squeezed coherent state for more details.

Both (photons and material) particles such as electrons create analogous interference patterns when passing through a double-slit experiment. For photons, this corresponds to the interference of a Maxwell light wave whereas, for material particles, this corresponds to the interference of the Schrödinger wave equation. Although this similarity might suggest that Maxwell's equations are simply Schrödinger's equation for photons, most physicists do not agree.[56][57] For one thing, they are mathematically different; most obviously, Schrödinger's one equation solves for a complex field, whereas Maxwell's four equations solve for real fields. More generally, the normal concept of a Schrödinger probability wave function cannot be applied to photons.[58] Being massless, they cannot be localized without being destroyed; technically, photons cannot have a position eigenstate $|\mathbf{r}\rangle$, and, thus, the normal Heisenberg uncertainty principle $\Delta x \Delta p > h/2$ does not pertain to photons. A few substitute wave functions have been suggested for the photon,[59][60][61][62] but they have not come into general use. Instead, physicists generally accept the second-quantized theory of photons described below, quantum electrodynamics, in which photons are quantized excitations of electromagnetic modes.

Another interpretation, that avoids duality, is the De Broglie–Bohm theory: known also as the *pilot-wave model*, the photon in this theory is both, wave and particle.[63] *"This idea seems to me so natural and simple, to resolve the wave-particle dilemma in such a clear and ordinary way, that it is a great mystery to me that it was so generally ignored"*,[64] J.S.Bell.

3.7 Bose–Einstein model of a photon gas

Main articles: Bose gas, Bose–Einstein statistics, Spin-statistics theorem and Gas in a box

In 1924, Satyendra Nath Bose derived Planck's law of black-body radiation without using any electromagnetism, but rather a modification of coarse-grained counting of phase space.[65] Einstein showed that this modification is equivalent to assuming that photons are rigorously identical and that it implied a "mysterious non-local interaction",[66][67] now understood as the requirement for a symmetric quantum mechanical state. This work led to the concept of coherent states and the development of the laser. In the same papers, Einstein extended Bose's formalism to material particles (bosons) and predicted that they would condense into their lowest quantum state at low enough temperatures; this Bose–Einstein condensation was observed experimentally in 1995.[68] It was later used by Lene Hau to slow, and then completely stop, light in 1999[69] and 2001.[70]

The modern view on this is that photons are, by virtue of their integer spin, bosons (as opposed to fermions with half-integer spin). By the spin-statistics theorem, all bosons obey Bose–Einstein statistics (whereas all fermions obey Fermi–Dirac statistics).[71]

3.8   Stimulated and spontaneous emission

Main articles: Stimulated emission and Laser

In 1916, Einstein showed that Planck's radiation law could be derived from a semi-classical, statistical treatment of photons and atoms, which implies a relation between the rates at which atoms emit and absorb photons. The condition follows from the assumption that light is emitted and absorbed by atoms independently, and that the thermal equilibrium is preserved by interaction with atoms. Consider a cavity in thermal equilibrium and filled with electromagnetic radiation and atoms that can emit and absorb that radiation. Thermal equilibrium requires that the energy density $\rho(\nu)$ of photons with frequency ν (which is proportional to their number density) is, on average, constant in time; hence, the rate at which photons of any particular frequency are *emitted* must equal the rate of *absorbing* them.[72]

Einstein began by postulating simple proportionality relations for the different reaction rates involved. In his model, the rate R_{ji} for a system to *absorb* a photon of frequency ν and transition from a lower energy E_j to a higher energy E_i is proportional to the number N_j of atoms with energy E_j and to the energy density $\rho(\nu)$ of ambient photons with that frequency,

$$R_{ji} = N_j B_{ji} \rho(\nu)$$

where B_{ji} is the rate constant for absorption. For the reverse process, there are two possibilities: spontaneous emission of a photon, and a return to the lower-energy state that is initiated by the interaction with a passing photon. Following Einstein's approach, the corresponding rate R_{ij} for the emission of photons of frequency ν and transition from a higher energy E_i to a lower energy E_j is

$$R_{ij} = N_i A_{ij} + N_i B_{ij} \rho(\nu)$$

where A_{ij} is the rate constant for emitting a photon spontaneously, and B_{ij} is the rate constant for emitting it in response to ambient photons (induced or stimulated emission). In thermodynamic equilibrium, the number of atoms in state i and that of atoms in state j must, on average, be constant; hence, the rates R_{ji} and R_{ij} must be equal. Also, by arguments analogous to the derivation of Boltzmann statistics, the ratio of N_i and N_j is $g_i/g_j \exp{(E_j - E_i)/kT}$, where $g_{i,j}$ are the degeneracy of the state i and that of j, respectively, $E_{i,j}$ their energies, k the Boltzmann constant and T the system's temperature. From this, it is readily derived that $g_i B_{ij} = g_j B_{ji}$ and

$$A_{ij} = \frac{8\pi h \nu^3}{c^3} B_{ij}.$$

The A and Bs are collectively known as the *Einstein coefficients*.[73]

Einstein could not fully justify his rate equations, but claimed that it should be possible to calculate the coefficients A_{ij}, B_{ji} and B_{ij} once physicists had obtained "mechanics and electrodynamics modified to accommodate the quantum hypothesis".[74] In fact, in 1926, Paul Dirac derived the B_{ij} rate constants in using a semiclassical approach,[75] and, in 1927, succeeded in deriving *all* the rate constants from first principles within the framework of quantum theory.[76][77] Dirac's work was the foundation of quantum electrodynamics, i.e., the quantization of the electromagnetic field itself. Dirac's approach is also called *second quantization* or quantum field theory;[78][79][80] earlier quantum mechanical treatments only treat material particles as quantum mechanical, not the electromagnetic field.

Einstein was troubled by the fact that his theory seemed incomplete, since it did not determine the *direction* of a spontaneously emitted photon. A probabilistic nature of light-particle motion was first considered by Newton in his treatment of birefringence and, more generally, of the splitting of light beams at interfaces into a transmitted beam and a reflected beam. Newton hypothesized that hidden variables in the light particle determined which path it would follow.[34] Similarly, Einstein hoped for a more complete theory that would leave nothing to chance, beginning his separation[45] from quantum mechanics. Ironically, Max Born's probabilistic interpretation of the wave function[81][82] was inspired by Einstein's later work searching for a more complete theory.[83]

3.9   Second quantization and high energy photon interactions

Main article: Quantum field theory

In 1910, Peter Debye derived Planck's law of black-body radiation from a relatively simple assumption.[84] He correctly decomposed the electromagnetic field in a cavity into its Fourier modes, and assumed that the energy in any mode was an integer multiple of $h\nu$, where ν is the frequency of the electromagnetic mode. Planck's law of black-body radiation follows immediately as a geometric sum. However, Debye's approach failed to give the correct formula for the energy fluctuations of blackbody radiation, which were derived by Einstein in 1909.[41]

In 1925, Born, Heisenberg and Jordan reinterpreted Debye's concept in a key way.[85] As may be shown classically, the Fourier modes of the electromagnetic field—a complete set of electromagnetic plane waves indexed by their wave vector **k** and polarization state—are equivalent to a set of uncoupled simple harmonic oscillators. Treated quantum mechanically, the energy levels of such oscillators are known to be $E = nh\nu$, where ν is the oscillator frequency. The key new step was to identify an electromagnetic mode with energy $E = nh\nu$ as a state with n photons, each of energy $h\nu$. This approach gives the correct energy fluctuation formula.

Dirac took this one step further.[76][77] He treated the interaction between a charge and an electromagnetic field as a small perturbation that induces transitions in the photon states, changing the numbers of photons in the modes, while conserving energy and momentum overall. Dirac was able to derive Einstein's A_{ij} and B_{ij} coefficients from first principles, and showed that the Bose–Einstein statistics of photons is a natural consequence of quantizing the electromagnetic field correctly (Bose's reasoning went in the opposite direction; he derived Planck's law of black-body radiation by *assuming* B–E statistics). In Dirac's time, it was not yet known that all bosons, including photons, must obey Bose–Einstein statistics.

Dirac's second-order perturbation theory can involve virtual photons, transient intermediate states of the electromagnetic field; the static electric and magnetic interactions are mediated by such virtual photons. In such quantum field theories, the probability amplitude of observable events is calculated by summing over *all* possible intermediate steps, even ones that are unphysical; hence, virtual photons are not constrained to satisfy $E = pc$, and may have extra polarization states; depending on the gauge used, virtual photons may have three or four polarization states, instead of the two states of real photons. Although these transient virtual photons can never be observed, they contribute measurably to the probabilities of observable events. Indeed, such second-order and higher-order perturbation calculations can give apparently infinite contributions to the sum. Such unphysical results are corrected for using the technique of renormalization.

Other virtual particles may contribute to the summation as well; for example, two photons may interact indirectly through virtual electron–positron pairs.[86] In fact, such photon-photon scattering (see two-photon physics), as well as electron-photon scattering, is meant to be one of the modes of operations of the planned particle accelerator, the International Linear Collider.[87]

In modern physics notation, the quantum state of the electromagnetic field is written as a Fock state, a tensor product of the states for each electromagnetic mode

$$|n_{k_0}\rangle \otimes |n_{k_1}\rangle \otimes \cdots \otimes |n_{k_n}\rangle \ldots$$

where $|n_{k_i}\rangle$ represents the state in which n_{k_i} photons are in the mode k_i. In this notation, the creation of a new photon in mode k_i (e.g., emitted from an atomic transition) is written as $|n_{k_i}\rangle \to |n_{k_i} + 1\rangle$. This notation merely expresses the concept of Born, Heisenberg and Jordan described above, and does not add any physics.

3.10   The hadronic properties of the photon

Measurements of the interaction between energetic photons and hadrons show that the interaction is much more intense than expected by the interaction of merely photons with the hadron's electric charge. Furthermore, the interaction of energetic photons with protons is similar to the interaction of photons with neutrons[88] in spite of the fact that the electric charge structures of protons and neutrons are substantially different.

A theory called Vector Meson Dominance (VMD) was developed to explain this effect. According to VMD, the photon is a superposition of the pure electromagnetic photon (which interacts only with electric charges) and vector meson.[89]

However, if experimentally probed at very short distances, the intrinsic structure of the photon is recognized as a flux of quark and gluon components, quasi-free according to asymptotic freedom in QCD and described by the photon structure function.[90][91] A comprehensive comparison of data with theoretical predictions is presented in a recent review.[92]

3.11   The photon as a gauge boson

Main article: Gauge theory

The electromagnetic field can be understood as a gauge field, i.e., as a field that results from requiring that a gauge symmetry holds independently at every position in spacetime.[93] For the electromagnetic field, this gauge symmetry is the Abelian U(1) symmetry of complex numbers of absolute value 1, which reflects the ability to vary the phase of a complex number without affecting observables or real valued functions made from it, such as the energy or the Lagrangian.

The quanta of an Abelian gauge field must be massless, uncharged bosons, as long as the symmetry is not broken; hence, the photon is predicted to be massless, and to have zero electric charge and integer spin. The particular form of the electromagnetic interaction specifies that the photon must have spin ± 1; thus, its helicity must be $\pm \hbar$. These two spin components correspond to the classical concepts of right-handed and left-handed circularly polarized light. However, the transient virtual photons of quantum electrodynamics may also adopt unphysical polarization states.[93]

In the prevailing Standard Model of physics, the photon is one of four gauge bosons in the electroweak interaction; the other three are denoted W^+, W^- and Z^0 and are responsible for the weak interaction. Unlike the photon, these gauge bosons have mass, owing to a mechanism that breaks their SU(2) gauge symmetry. The unification of the photon with W and Z gauge bosons in the electroweak interaction was accomplished by Sheldon Glashow, Abdus Salam and Steven Weinberg, for which they were awarded the 1979 Nobel Prize in physics.[94][95][96] Physicists continue to hypothesize grand unified theories that connect these four gauge bosons with the eight gluon gauge bosons of quantum chromodynamics; however, key predictions of these theories, such as proton decay, have not been observed experimentally.[97]

3.12   Contributions to the mass of a system

See also: Mass in special relativity and General relativity

The energy of a system that emits a photon is *decreased* by the energy E of the photon as measured in the rest frame of the emitting system, which may result in a reduction in mass in the amount E/c^2. Similarly, the mass of a system that absorbs a photon is *increased* by a corresponding amount. As an application, the energy balance of nuclear reactions involving photons is commonly written in terms of the masses of the nuclei involved, and terms of the form E/c^2 for the gamma photons (and for other relevant energies, such as the recoil energy of nuclei).[98]

This concept is applied in key predictions of quantum electrodynamics (QED, see above). In that theory, the mass of electrons (or, more generally, leptons) is modified by including the mass contributions of virtual photons, in a technique known as renormalization. Such "radiative corrections" contribute to a number of predictions of QED, such as the magnetic dipole moment of leptons, the Lamb shift, and the hyperfine structure of bound lepton pairs, such as muonium and positronium.[99]

Since photons contribute to the stress–energy tensor, they exert a gravitational attraction on other objects, according to the theory of general relativity. Conversely, photons are themselves affected by gravity; their normally straight trajectories may be bent by warped spacetime, as in gravitational lensing, and their frequencies may be lowered by moving to a higher gravitational potential, as in the Pound–Rebka experiment. However, these effects are not specific to photons; exactly the same effects would be predicted for classical electromagnetic waves.[100]

3.13 Photons in matter

See also: Group velocity and Photochemistry

Any 'explanation' of how photons travel through matter has to explain why different arrangements of matter are transparent or opaque at different wavelengths (light through carbon as diamond or not, as graphite) and why individual photons behave in the same way as large groups. Explanations that invoke 'absorption' and 're-emission' have to provide an explanation for the directionality of the photons (diffraction, reflection) and further explain how entangled photon pairs can travel through matter without their quantum state collapsing.

The simplest explanation is that light that travels through transparent matter does so at a lower speed than c, the speed of light in a vacuum. In addition, light can also undergo scattering and absorption. There are circumstances in which heat transfer through a material is mostly radiative, involving emission and absorption of photons within it. An example would be in the core of the Sun. Energy can take about a million years to reach the surface.[101] However, this phenomenon is distinct from scattered radiation passing diffusely through matter, as it involves local equilibrium between the radiation and the temperature. Thus, the time is how long it takes the *energy* to be transferred, not the *photons* themselves. Once in open space, a photon from the Sun takes only 8.3 minutes to reach Earth. The factor by which the speed of light is decreased in a material is called the refractive index of the material. In a classical wave picture, the slowing can be explained by the light inducing electric polarization in the matter, the polarized matter radiating new light, and the new light interfering with the original light wave to form a delayed wave. In a particle picture, the slowing can instead be described as a blending of the photon with quantum excitation of the matter (quasi-particles such as phonons and excitons) to form a polariton; this polariton has a nonzero effective mass, which means that it cannot travel at c.

Alternatively, photons may be viewed as *always* traveling at c, even in matter, but they have their phase shifted (delayed or advanced) upon interaction with atomic scatters: this modifies their wavelength and momentum, but not speed.[102] A light wave made up of these photons does travel slower than the speed of light. In this view the photons are "bare", and are scattered and phase shifted, while in the view of the preceding paragraph the photons are "dressed" by their interaction with matter, and move without scattering or phase shifting, but at a lower speed.

Light of different frequencies may travel through matter at different speeds; this is called dispersion. In some cases, it can result in extremely slow speeds of light in matter. The effects of photon interactions with other quasi-particles may be observed directly in Raman scattering and Brillouin scattering.[103]

Photons can also be absorbed by nuclei, atoms or molecules, provoking transitions between their energy levels. A classic example is the molecular transition of retinal $C_{20}H_{28}O$, which is responsible for vision, as discovered in 1958 by Nobel laureate biochemist George Wald and co-workers. The absorption provokes a cis-trans isomerization that, in combination with other such transitions, is transduced into nerve impulses. The absorption of photons can even break chemical bonds, as in the photodissociation of chlorine; this is the subject of photochemistry.[104][105] Analogously, gamma rays can in some circumstances dissociate atomic nuclei in a process called photodisintegration.

3.14   Technological applications

Photons have many applications in technology. These examples are chosen to illustrate applications of photons *per se*, rather than general optical devices such as lenses, etc. that could operate under a classical theory of light. The laser is an extremely important application and is discussed above under stimulated emission.

Individual photons can be detected by several methods. The classic photomultiplier tube exploits the photoelectric effect: a photon landing on a metal plate ejects an electron, initiating an ever-amplifying avalanche of electrons. Charge-coupled device chips use a similar effect in semiconductors: an incident photon generates a charge on a microscopic capacitor that can be detected. Other detectors such as Geiger counters use the ability of photons to ionize gas molecules, causing a detectable change in conductivity.[106]

Planck's energy formula $E = h\nu$ is often used by engineers and chemists in design, both to compute the change in energy resulting from a photon absorption and to predict the frequency of the light emitted for a given energy transition. For example, the emission spectrum of a fluorescent light bulb can be designed using gas molecules with different electronic energy levels and adjusting the typical energy with which an electron hits the gas molecules within the bulb.[Note 6]

Under some conditions, an energy transition can be excited by "two" photons that individually would be insufficient. This allows for higher resolution microscopy, because the sample absorbs energy only in the region where two beams of different colors overlap significantly, which can be made much smaller than the excitation volume of a single beam (see two-photon excitation microscopy). Moreover, these photons cause less damage to the sample, since they are of lower energy.[107]

In some cases, two energy transitions can be coupled so that, as one system absorbs a photon, another nearby system "steals" its energy and re-emits a photon of a different frequency. This is the basis of fluorescence resonance energy transfer, a technique that is used in molecular biology to study the interaction of suitable proteins.[108]

Several different kinds of hardware random number generator involve the detection of single photons. In one example, for each bit in the random sequence that is to be produced, a photon is sent to a beam-splitter. In such a situation, there are two possible outcomes of equal probability. The actual outcome is used to determine whether the next bit in the sequence is "0" or "1".[109][110]

3.15   Recent research

See also: Quantum optics

Much research has been devoted to applications of photons in the field of quantum optics. Photons seem well-suited to be elements of an extremely fast quantum computer, and the quantum entanglement of photons is a focus of research. Nonlinear optical processes are another active research area, with topics such as two-photon absorption, self-phase modulation, modulational instability and optical parametric oscillators. However, such processes generally do not require the assumption of photons *per se*; they may often be modeled by treating atoms as nonlinear oscillators. The nonlinear process of spontaneous parametric down conversion is often used to produce single-photon states. Finally, photons are essential in some aspects of optical communication, especially for quantum cryptography.[Note 7]

3.16   See also

- Advanced Photon Source at Argonne National Laboratory
- Ballistic photon
- Doppler shift
- Electromagnetic radiation
- HEXITEC

- Laser

- Light

- Luminiferous aether

- Medipix

- Phonons

- Photon counting

- Photon energy

- Photon polarization

- Photonic molecule

- Photography

- Photonics

- Quantum optics

- Single photon sources

- Static forces and virtual-particle exchange

- Two-photon physics

- EPR paradox

- Dirac equation

3.17 Notes

[1] Although the 1967 Elsevier translation of Planck's Nobel Lecture interprets Planck's *Lichtquant* as "photon", the more literal 1922 translation by Hans Thacher Clarke and Ludwik Silberstein *The origin and development of the quantum theory*, The Clarendon Press, 1922 (here) uses "light-quantum". No evidence is known that Planck himself used the term "photon" by 1926 (see also this note).

[2] Isaac Asimov credits Arthur Compton with defining quanta of energy as photons in 1923. Asimov, I. (1966). *The Neutrino, Ghost Particle of the Atom*. Garden City (NY): Doubleday. ISBN 0-380-00483-6. LCCN 66017073. and Asimov, I. (1966). *The Universe From Flat Earth To Quasar*. New York (NY): Walker. ISBN 0-8027-0316-X. LCCN 66022515.

[3] The mass of the photon is believed to be exactly zero, based on experiment and theoretical considerations described in the article. Some sources also refer to the *relativistic mass* concept, which is just the energy scaled to units of mass. For a photon with wavelength λ or energy E, this is $h/\lambda c$ or E/c^2. This usage for the term "mass" is no longer common in scientific literature. Further info: What is the mass of a photon? http://math.ucr.edu/home/baez/physics/ParticleAndNuclear/photon_mass.html

[4] The phrase "no matter how intense" refers to intensities below approximately 10^{13} W/cm^2 at which point perturbation theory begins to break down. In contrast, in the intense regime, which for visible light is above approximately 10^{14} W/cm^2, the classical wave description correctly predicts the energy acquired by electrons, called ponderomotive energy. (See also: Boreham *et al.* (1996). "Photon density and the correspondence principle of electromagnetic interaction".) By comparison, sunlight is only about 0.1 W/cm^2.

[5] These experiments produce results that cannot be explained by any classical theory of light, since they involve anticorrelations that result from the quantum measurement process. In 1974, the first such experiment was carried out by Clauser, who reported a violation of a classical Cauchy–Schwarz inequality. In 1977, Kimble *et al.* demonstrated an analogous anti-bunching effect of photons interacting with a beam splitter; this approach was simplified and sources of error eliminated in the photon-anticorrelation experiment of Grangier *et al.* (1986). This work is reviewed and simplified further in Thorn *et al.* (2004). (These references are listed below under #Additional references.)

[6] An example is US Patent Nr. 5212709.

[7] Introductory-level material on the various sub-fields of quantum optics can be found in Fox, M. (2006). *Quantum Optics: An Introduction*. Oxford University Press. ISBN 0-19-856673-5.

3.18 References

[1] Amsler, C. (Particle Data Group); Amsler; Doser; Antonelli; Asner; Babu; Baer; Band; Barnett; Bergren; Beringer; Bernardi; Bertl; Bichsel; Biebel; Bloch; Blucher; Blusk; Cahn; Carena; Caso; Ceccucci; Chakraborty; Chen; Chivukula; Cowan; Dahl; d'Ambrosio; Damour; et al. (2008). "Review of Particle Physics: Gauge and Higgs bosons" (PDF). *Physics Letters B* **667**: 1. Bibcode:2008PhLB..667....1P. doi:10.1016/j.physletb.2008.07.018.

[2] Joos, George (1951). *Theoretical Physics*. London and Glasgow: Blackie and Son Limited. p. 679.

[3] Kimble, H.J.; Dagenais, M.; Mandel, L.; Dagenais; Mandel (1977). "Photon Anti-bunching in Resonance Fluorescence". *Physical Review Letters* **39** (11): 691–695. Bibcode:1977PhRvL..39..691K. doi:10.1103/PhysRevLett.39.691.

[4] Grangier, P.; Roger, G.; Aspect, A.; Roger; Aspect (1986). "Experimental Evidence for a Photon Anticorrelation Effect on a Beam Splitter: A New Light on Single-Photon Interferences". *Europhysics Letters* **1** (4): 173–179. Bibcode:1986EL......1..173G. doi:10.1209/0295-5075/1/4/004.

[5] Planck, M. (1901). "On the Law of Distribution of Energy in the Normal Spectrum". *Annalen der Physik* **4** (3): 553–563. Bibcode:1901AnP...309..553P. doi:10.1002/andp.19013090310. Archived from the original on 2008-04-18.

[6] Einstein, A. (1905). "Über einen die Erzeugung und Verwandlung des Lichtes betreffenden heuristischen Gesichtspunkt" (PDF). *Annalen der Physik* (in German) **17** (6): 132–148. Bibcode:1905AnP...322..132E. doi:10.1002/andp.19053220607.. An English translation is available from Wikisource.

[7] "Discordances entre l'expérience et la théorie électromagnétique du rayonnement." In Électrons et Photons. Rapports et Discussions de Cinquième Conseil de Physique, edited by Institut International de Physique Solvay. Paris: Gauthier-Villars, pp. 55-85.

[8] Helge Kragh: *Photon: New light on an old name*. Arxiv, 2014-2-28

[9] Villard, P. (1900). "Sur la réflexion et la réfraction des rayons cathodiques et des rayons déviables du radium". *Comptes Rendus des Séances de l'Académie des Sciences* (in French) **130**: 1010–1012.

[10] Villard, P. (1900). "Sur le rayonnement du radium". *Comptes Rendus des Séances de l'Académie des Sciences* (in French) **130**: 1178–1179.

[11] Rutherford, E.; Andrade, E.N.C. (1914). "The Wavelength of the Soft Gamma Rays from Radium B". *Philosophical Magazine* **27** (161): 854–868. doi:10.1080/14786440508635156.

[12] Kobychev, V.V.; Popov, S.B. (2005). "Constraints on the photon charge from observations of extragalactic sources". *Astronomy Letters* **31** (3): 147–151. arXiv:hep-ph/0411398. Bibcode:2005AstL...31..147K. doi:10.1134/1.1883345.

[13] Role as gauge boson and polarization section 5.1 inAitchison, I.J.R.; Hey, A.J.G. (1993). *Gauge Theories in Particle Physics*. IOP Publishing. ISBN 0-85274-328-9.

[14] See p.31 inAmsler, C.; et al. (2008). "Review of Particle Physics". *Physics Letters B* **667**: 1–1340. Bibcode:2008PhLB..667....1P. doi:10.1016/j.physletb.2008.07.018.

[15] Halliday, David; Resnick, Robert; Walker, Jerl (2005), *Fundamental of Physics* (7th ed.), USA: John Wiley and Sons, Inc., ISBN 0-471-23231-9

[16] See section 1.6 in Alonso, M.; Finn, E.J. (1968). *Fundamental University Physics Volume III: Quantum and Statistical Physics*. Addison-Wesley. ISBN 0-201-00262-0.

[17] Davison E. Soper, Electromagnetic radiation is made of photons, Institute of Theoretical Science, University of Oregon

[18] This property was experimentally verified by Raman and Bhagavantam in 1931: Raman, C.V.; Bhagavantam, S. (1931). "Experimental proof of the spin of the photon" (PDF). *Indian Journal of Physics* **6**: 353.

[19] Burgess, C.; Moore, G. (2007). "1.3.3.2". *The Standard Model. A Primer*. Cambridge University Press. ISBN 0-521-86036-9.

[20] Griffiths, David J. (2008), *Introduction to Elementary Particles* (2nd revised ed.), WILEY-VCH, ISBN 978-3-527-40601-2

[21] E.g., section 9.3 in Alonso, M.; Finn, E.J. (1968). *Fundamental University Physics Volume III: Quantum and Statistical Physics*. Addison-Wesley.

[22] E.g., Appendix XXXII in Born, M. (1962). *Atomic Physics*. Blackie & Son. ISBN 0-486-65984-4.

[23] Mermin,David(February1984). "Relativity without light".*American Journal of Physics***52**(2): 119–124. Bibcode:1984AmJPh... doi:10.1119/1.13917.

[24] Plimpton, S.; Lawton, W. (1936). "A Very Accurate Test of Coulomb's Law of Force Between Charges". *Physical Review* **50** (11): 1066. Bibcode:1936PhRv...50.1066P. doi:10.1103/PhysRev.50.1066.

[25] Williams, E.; Faller, J.; Hill, H. (1971). "New Experimental Test of Coulomb's Law: A Laboratory Upper Limit on the Photon Rest Mass". *Physical Review Letters* **26** (12): 721. Bibcode:1971PhRvL..26..721W. doi:10.1103/PhysRevLett.26.721.

[26] Chibisov,G V(1976). "Astrophysical upper limits on the photon rest mass".*Soviet Physics Uspekhi***19**(7): 624. Bibcode:1976. doi:10.1070/PU1976v019n07ABEH005277.

[27] Lakes, Roderic (1998). "Experimental Limits on the Photon Mass and Cosmic Magnetic Vector Potential". *Physical Review Letters* **80** (9): 1826. Bibcode:1998PhRvL..80.1826L. doi:10.1103/PhysRevLett.80.1826.

[28] Amsler, C; Doser, M; Antonelli, M; Asner, D; Babu, K; Baer, H; Band, H; Barnett, R; et al. (2008). "Review of Particle Physics∗". *Physics Letters B* **667**: 1. Bibcode:2008PhLB..667....1P. doi:10.1016/j.physletb.2008.07.018. Summary Table

[29] Adelberger, Eric; Dvali, Gia; Gruzinov, Andrei (2007). "Photon-Mass Bound Destroyed by Vortices". *Physical Review Letters* **98** (1): 010402. arXiv:hep-ph/0306245. Bibcode:2007PhRvL..98a0402A. doi:10.1103/PhysRevLett.98.010402. PMID 17358459. preprint

[30] Wilczek, Frank (2010). *The Lightness of Being: Mass, Ether, and the Unification of Forces*. Basic Books. p. 212. ISBN 978-0-465-01895-6.

[31] Descartes, R. (1637). *Discours de la méthode (Discourse on Method)* (in French). Imprimerie de Ian Maire. ISBN 0-268-00870-1.

[32] Hooke, R. (1667). *Micrographia: or some physiological descriptions of minute bodies made by magnifying glasses with observations and inquiries thereupon ...* London (UK): Royal Society of London. ISBN 0-486-49564-7.

[33] Huygens, C. (1678). *Traité de la lumière* (in French).. An English translation is available from Project Gutenberg

[34] Newton, I. (1952) [1730]. *Opticks* (4th ed.). Dover (NY): Dover Publications. Book II, Part III, Propositions XII–XX; Queries 25–29. ISBN 0-486-60205-2.

[35] Buchwald, J.Z. (1989). *The Rise of the Wave Theory of Light: Optical Theory and Experiment in the Early Nineteenth Century*. University of Chicago Press. ISBN 0-226-07886-8. OCLC 18069573.

[36] Maxwell, J.C. (1865). "A Dynamical Theory of the Electromagnetic Field". *Philosophical Transactions of the Royal Society* **155**: 459–512. Bibcode:1865RSPT..155..459C. doi:10.1098/rstl.1865.0008. This article followed a presentation by Maxwell on 8 December 1864 to the Royal Society.

[37] Hertz, H. (1888). "Über Strahlen elektrischer Kraft". *Sitzungsberichte der Preussischen Akademie der Wissenschaften (Berlin)* (in German) **1888**: 1297–1307.

[38] Frequency-dependence of luminiscence p. 276f., photoelectric effect section 1.4 in Alonso, M.; Finn, E.J. (1968). *Fundamental University Physics Volume III: Quantum and Statistical Physics*. Addison-Wesley. ISBN 0-201-00262-0.

[39] Wien, W. (1911). "Wilhelm Wien Nobel Lecture".

[40] Planck, M. (1920). "Max Planck's Nobel Lecture".

[41] Einstein, A. (1909). "Über die Entwicklung unserer Anschauungen über das Wesen und die Konstitution der Strahlung" (PDF). *Physikalische Zeitschrift* (in German) **10**: 817–825.. An English translation is available from Wikisource.

[42] Presentation speech by Svante Arrhenius for the 1921 Nobel Prize in Physics, December 10, 1922. Online text from [nobel-prize.org], The Nobel Foundation 2008. Access date 2008-12-05.

[43] Einstein, A. (1916). "Zur Quantentheorie der Strahlung". *Mitteilungen der Physikalischen Gesellschaft zu Zürich* **16**: 47. Also *Physikalische Zeitschrift*, **18**, 121–128 (1917). (German)

[44] Compton, A. (1923). "A Quantum Theory of the Scattering of X-rays by Light Elements". *Physical Review* **21** (5): 483–502. Bibcode:1923PhRv...21..483C. doi:10.1103/PhysRev.21.483.

[45] Pais, A. (1982). *Subtle is the Lord: The Science and the Life of Albert Einstein*. Oxford University Press. ISBN 0-19-853907-X.

[46] *Einstein and the Quantum: The Quest of the Valiant Swabian*, A. Douglas Stone, Princeton University Press, 2013.

[47] Millikan, R.A (1924). "Robert A. Millikan's Nobel Lecture".

[48] Hendry, J. (1980). "The development of attitudes to the wave-particle duality of light and quantum theory, 1900–1920". *Annals of Science* **37** (1): 59–79. doi:10.1080/00033798000200121.

[49] Bohr, N.; Kramers, H.A.; Slater, J.C. (1924). "The Quantum Theory of Radiation". *Philosophical Magazine* **47**: 785–802. doi:10.1080/14786442408565262. Also *Zeitschrift für Physik*, **24**, 69 (1924).

[50] Heisenberg, W. (1933). "Heisenberg Nobel lecture".

[51] Mandel, L. (1976). E. Wolf, ed. "The case for and against semiclassical radiation theory". *Progress in Optics*. Progress in Optics (North-Holland) **13**: 27–69. doi:10.1016/S0079-6638(08)70018-0. ISBN 978-0-444-10806-7.

[52] Taylor, G.I. (1909). *Interference fringes with feeble light*. Proceedings of the Cambridge Philosophical Society **15**: 114–115.

[53] Saleh, B. E. A. and Teich, M. C. (2007). *Fundamentals of Photonics*. Wiley. ISBN 0-471-35832-0.

[54] Heisenberg, W. (1927). "Über den anschaulichen Inhalt der quantentheoretischen Kinematik und Mechanik". *Zeitschrift für Physik* (in German) **43** (3–4): 172–198. Bibcode:1927ZPhy...43..172H. doi:10.1007/BF01397280.

[55] E.g., p. 10f. in Schiff, L.I. (1968). *Quantum Mechanics* (3rd ed.). McGraw-Hill. ASIN B001B3MINM. ISBN 0-07-055287-8.

[56] Kramers, H.A. (1958). *Quantum Mechanics*. Amsterdam: North-Holland. ASIN B0006AUW5C. ISBN 0-486-49533-7.

[57] Bohm, D. (1989) [1954]. *Quantum Theory*. Dover Publications. ISBN 0-486-65969-0.

[58] Newton, T.D.; Wigner, E.P. (1949). "Localized states for elementary particles". *Reviews of Modern Physics* **21** (3): 400–406. Bibcode:1949RvMP...21..400N. doi:10.1103/RevModPhys.21.400.

[59] Bialynicki-Birula, I. (1994). "On the wave function of the photon" (PDF). *Acta Physica Polonica A* **86**: 97–116.

[60] Sipe,J.E. (1995). "Photon wave functions".*Physical Review A* **52**(3): 1875–1883. Bibcode:1995PhRvA..52.1875S.doi:10.875.

[61] Bialynicki-Birula, I. (1996). "Photon wave function". *Progress in Optics*. Progress in Optics **36**: 245–294. doi:10.1016/S0079-6638(08)70316-0. ISBN 978-0-444-82530-8.

[62] Scully, M.O.; Zubairy, M.S. (1997). *Quantum Optics*. Cambridge (UK): Cambridge University Press. ISBN 0-521-43595-1.

[63] The best illustration is the Couder experiment, demonstrating the behaviour of a mechanical analog, see https://www.youtube.com/watch?v=W9yWv5dqSKk

[64] Bell, J. S., "Speakable and Unspeakable in Quantum Mechanics", Cambridge: Cambridge University Press, 1987.

[65] Bose,S.N.(1924). "Plancks Gesetz und Lichtquantenhypothese".*Zeitschrift für Physik*(in German)**26**: 178–181. Bibcode:B. doi:10.1007/BF01327326.

[66] Einstein, A. (1924). "Quantentheorie des einatomigen idealen Gases". *Sitzungsberichte der Preussischen Akademie der Wissenschaften (Berlin), Physikalisch-mathematische Klasse* (in German) **1924**: 261–267.

[67] Einstein, A. (1925). "Quantentheorie des einatomigen idealen Gases, Zweite Abhandlung". *Sitzungsberichte der Preussischen Akademie der Wissenschaften (Berlin), Physikalisch-mathematische Klasse* (in German) **1925**: 3–14. doi:10.1002/3527608958.ch28. ISBN 978-3-527-60895-9.

[68] Anderson, M.H.; Ensher, J.R.; Matthews, M.R.; Wieman, C.E.; Cornell, E.A. (1995). "Observation of Bose–Einstein Condensation in a Dilute Atomic Vapor". *Science* **269** (5221): 198–201. Bibcode:1995Sci...269..198A. doi:10.1126/science.269.5221.. JSTOR2888436. PMID 17789847.

[69] "Physicists Slow Speed of Light". News.harvard.edu (1999-02-18). Retrieved on 2015-05-11.

[70] "Light Changed to Matter, Then Stopped and Moved". photonics.com (February 2007). Retrieved on 2015-05-11.

[71] Streater, R.F.; Wightman, A.S. (1989). *PCT, Spin and Statistics, and All That*. Addison-Wesley. ISBN 0-201-09410-X.

[72] Einstein, A. (1916). "Strahlungs-emission und -absorption nach der Quantentheorie". *Verhandlungen der Deutschen Physikalischen Gesellschaft* (in German) **18**: 318–323. Bibcode:1916DPhyG..18..318E.

[73] Section 1.4 in Wilson, J.; Hawkes, F.J.B. (1987). *Lasers: Principles and Applications*. New York: Prentice Hall. ISBN 0-13-523705-X.

[74] P. 322 in Einstein, A. (1916). "Strahlungs-emission und -absorption nach der Quantentheorie". *Verhandlungen der Deutschen Physikalischen Gesellschaft* (in German) **18**: 318–323. Bibcode:1916DPhyG..18..318E.:

> Die Konstanten A_m^n and B_m^n würden sich direkt berechnen lassen, wenn wir im Besitz einer im Sinne der Quantenhypothese modifizierten Elektrodynamik und Mechanik wären."

[75] Dirac, P.A.M. (1926). "On the Theory of Quantum Mechanics". *Proceedings of the Royal Society A* **112** (762): 661–677. Bibcode:1926RSPSA.112..661D. doi:10.1098/rspa.1926.0133.

[76] Dirac, P.A.M. (1927). "The Quantum Theory of the Emission and Absorption of Radiation" (PDF). *Proceedings of the Royal Society A* **114** (767): 243–265. Bibcode:1927RSPSA.114..243D. doi:10.1098/rspa.1927.0039.

[77] Dirac, P.A.M. (1927b). *The Quantum Theory of Dispersion. Proceedings of the Royal Society A* **114**: 710–728. doi:10.1098/rspa.1.

[78] Heisenberg, W.; Pauli, W. (1929). "Zur Quantentheorie der Wellenfelder". *Zeitschrift für Physik* (in German) **56**: 1. Bibcode:192H. doi:10.1007/BF01340129.

[79] Heisenberg, W.; Pauli, W. (1930). "Zur Quantentheorie der Wellenfelder". *Zeitschrift für Physik* (in German) **59** (3–4): 139. Bibcode:1930ZPhy...59..168H. doi:10.1007/BF01341423.

[80] Fermi, E. (1932). "Quantum Theory of Radiation" (PDF). *Reviews of Modern Physics* **4**: 87. Bibcode:1932RvMP....4...87F. doi:10.1103/RevModPhys.4.87.

[81] Born, M. (1926). "Zur Quantenmechanik der Stossvorgänge" (PDF). *Zeitschrift für Physik* (in German) **37** (12): 863–867. Bibcode:1926ZPhy...37..863B. doi:10.1007/BF01397477.

[82] Born, M. (1926). "Quantenmechanik der Stossvorgänge". *Zeitschrift für Physik* (in German) **38** (11–12): 803. Bibcode:1926ZPhy. doi:10.1007/BF01397184.

[83] Pais, A. (1986). *Inward Bound: Of Matter and Forces in the Physical World*. Oxford University Press. p. 260. ISBN 0-19-851997-4. Specifically, Born claimed to have been inspired by Einstein's never-published attempts to develop a "ghost-field" theory, in which point-like photons are guided probabilistically by ghost fields that follow Maxwell's equations.

[84] Debye, P. (1910). "Der Wahrscheinlichkeitsbegriff in der Theorie der Strahlung". *Annalen der Physik* (in German) **33** (16): 1427–1434. Bibcode:1910AnP...338.1427D. doi:10.1002/andp.19103381617.

[85] Born, M.; Heisenberg, W.; Jordan, P. (1925). "Quantenmechanik II". *Zeitschrift für Physik* (in German) **35** (8–9): 557–615. Bibcode:1926ZPhy...35..557B. doi:10.1007/BF01379806.

[86] Photon-photon-scattering section 7-3-1, renormalization chapter 8-2 in Itzykson, C.; Zuber, J.-B. (1980). *Quantum Field Theory*. McGraw-Hill. ISBN 0-07-032071-3.

[87] Weiglein, G. (2008). "Electroweak Physics at the ILC". *Journal of Physics: Conference Series* **110** (4): 042033. arXiv:0711.3003. Bibcode:2008JPhCS.110d2033W. doi:10.1088/1742-6596/110/4/042033.

[88] Bauer, T. H.; Spital, R. D.; Yennie, D. R.; Pipkin, F. M. (1978). "The hadronic properties of the photon in high-energy interactions". *Reviews of Modern Physics* **50** (2): 261. Bibcode:1978RvMP...50..261B. doi:10.1103/RevModPhys.50.261.

[89] Sakurai, J. J. (1960). "Theory of strong interactions". *Annals of Physics* **11**: 1. Bibcode:1960AnPhy..11....1S. doi:10.1016/0003-4916(60)90126-3.

[90] Walsh,T.F.;Zerwas,P. (1973). "Two-photon processes in the parton model".*Physics Letters B***44**(2): 195. Bibcode:1973PhLBW. doi:10.1016/0370-2693(73)90520-0.

[91] Witten, E. (1977). "Anomalous cross section for photon-photon scattering in gauge theories". *Nuclear Physics B* **120** (2): 189. Bibcode:1977NuPhB.120..189W. doi:10.1016/0550-3213(77)90038-4.

[92] Nisius, R. (2000). "The photon structure from deep inelastic electron–photon scattering". *Physics Reports* **332** (4–6): 165. Bibcode:2000PhR...332..165N. doi:10.1016/S0370-1573(99)00115-5.

[93] Ryder, L.H. (1996). *Quantum field theory* (2nd ed.). Cambridge University Press. ISBN 0-521-47814-6.

[94] Sheldon Glashow Nobel lecture, delivered 8 December 1979.

[95] Abdus Salam Nobel lecture, delivered 8 December 1979.

[96] Steven Weinberg Nobel lecture, delivered 8 December 1979.

[97] E.g., chapter 14 in Hughes, I. S. (1985). *Elementary particles* (2nd ed.). Cambridge University Press. ISBN 0-521-26092-2.

[98] E.g., section 10.1 in Dunlap, R.A. (2004).*An Introduction to the Physics of Nuclei and Particles*. Brooks/Cole. ISBN 0-53-39294-6.

[99] Radiative correction to electron mass section 7-1-2, anomalous magnetic moments section 7-2-1, Lamb shift section 7-3-2 and hyperfine splitting in positronium section 10-3 in Itzykson, C.; Zuber, J.-B. (1980). *Quantum Field Theory*. McGraw-Hill. ISBN 0-07-032071-3.

[100] E. g. sections 9.1 (gravitational contribution of photons) and 10.5 (influence of gravity on light) in Stephani, H.; Stewart, J. (1990). *General Relativity: An Introduction to the Theory of Gravitational Field*. Cambridge University Press. pp. 86 ff, 108 ff. ISBN 0-521-37941-5.

[101] Naeye, R. (1998). *Through the Eyes of Hubble: Birth, Life and Violent Death of Stars*. CRC Press. ISBN 0-7503-0484-7. OCLC 40180195.

[102] Ch 4 in Hecht, Eugene (2001). *Optics*. Addison Wesley. ISBN 978-0-8053-8566-3.

[103] Polaritons section 10.10.1, Raman and Brillouin scattering section 10.11.3 in Patterson, J.D.; Bailey, B.C. (2007). *Solid-State Physics: Introduction to the Theory*. Springer. pp. 569 ff, 580 ff. ISBN 3-540-24115-9.

[104] E.g., section 11-5 C in Pine, S.H.; Hendrickson, J.B.; Cram, D.J.; Hammond, G.S. (1980). *Organic Chemistry* (4th ed.). McGraw-Hill. ISBN 0-07-050115-7.

[105] Nobel lecture given by G. Wald on December 12, 1967, online at nobelprize.org: The Molecular Basis of Visual Excitation.

[106] Photomultiplier section 1.1.10, CCDs section 1.1.8, Geiger counters section 1.3.2.1 in Kitchin, C.R. (2008). *Astrophysical Techniques*. Boca Raton (FL): CRC Press. ISBN 1-4200-8243-4.

[107] Denk, W.; Svoboda, K. (1997). "Photon upmanship: Why multiphoton imaging is more than a gimmick". *Neuron* **18** (3): 351–357. doi:10.1016/S0896-6273(00)81237-4. PMID 9115730.

[108] Lakowicz, J.R. (2006). *Principles of Fluorescence Spectroscopy*. Springer. pp. 529 ff. ISBN 0-387-31278-1.

[109] Jennewein, T.; Achleitner, U.; Weihs, G.; Weinfurter, H.; Zeilinger, A. (2000). "A fast and compact quantum random number generator". *Review of Scientific Instruments* **71** (4): 1675–1680. arXiv:quant-ph/9912118. Bibcode:2000RScI...71.1675J. doi:10.1063/1.1150518.

[110] Stefanov, A.; Gisin, N.; Guinnard, O.; Guinnard, L.; Zbiden, H. (2000). "Optical quantum random number generator". *Journal of Modern Optics* **47** (4): 595–598. doi:10.1080/095003400147908.

3.19 Additional references

By date of publication:

- Clauser, J.F. (1974). "Experimental distinction between the quantum and classical field-theoretic predictions for the photoelectric effect".*Physical Review D***9**(4): 853–860. Bibcode:1974PhRvD...9..853C.doi:10.1103/PhysRevD.9.

- Kimble, H.J.; Dagenais, M.; Mandel, L. (1977). "Photon Anti-bunching in Resonance Fluorescence". *Physical Review Letters* **39** (11): 691–695. Bibcode:1977PhRvL..39..691K. doi:10.1103/PhysRevLett.39.691.

- Pais, A. (1982). *Subtle is the Lord: The Science and the Life of Albert Einstein.* Oxford University Press.

- Feynman, Richard (1985). *QED: The Strange Theory of Light and Matter.* Princeton University Press. ISBN 978-0-691-12575-6.

- Grangier, P.; Roger, G.; Aspect, A. (1986). "Experimental Evidence for a Photon Anticorrelation Effect on a Beam Splitter: A New Light on Single-Photon Interferences". *Europhysics Letters* **1** (4): 173–179. Bibcode:1986EL .doi:10.1209/0295-5075/1/4/004.

- Lamb,W.E.(1995). "Anti-photon".*Applied Physics B***60**(2–3): 77–84. Bibcode:1995ApPhB..60...77L.doi:16.

- Special supplemental issue of *Optics and Photonics News* (vol. 14, October 2003) article web link

  - Roychoudhuri, C.; Rajarshi, R. (2003). "The nature of light: what is a photon?". *Optics and Photonics News* **14**: S1 (Supplement).

  - Zajonc, A. "Light reconsidered". *Optics and Photonics News* **14**: S2–S5 (Supplement).

  - Loudon, R. "What is a photon?". *Optics and Photonics News* **14**: S6–S11 (Supplement).

  - Finkelstein, D. "What is a photon?". *Optics and Photonics News* **14**: S12–S17 (Supplement).

  - Muthukrishnan, A.; Scully, M.O.; Zubairy, M.S. "The concept of the photon—revisited". *Optics and Photonics News* **14**: S18–S27 (Supplement).

  - Mack, H.; Schleich, W.P.. "A photon viewed from Wigner phase space". *Optics and Photonics News* **14**: S28–S35 (Supplement)

- Glauber, R. (2005). "One Hundred Years of Light Quanta" (PDF). *2005 Physics Nobel Prize Lecture.*

- Hentschel, K. (2007). "Light quanta: The maturing of a concept by the stepwise accretion of meaning". *Physics and Philosophy* **1** (2): 1–20.

Education with single photons:

- Thorn, J.J.; Neel, M.S.; Donato, V.W.; Bergreen, G.S.; Davies, R.E.; Beck, M. (2004). "Observing the quantum behavior of light in an undergraduate laboratory" (PDF). *American Journal of Physics* **72** (9): 1210–1219. Bibcode:2004AmJPh..72.1210T. doi:10.1119/1.1737397.

- Bronner, P.; Strunz, Andreas; Silberhorn, Christine; Meyn, Jan-Peter (2009). "Interactive screen experiments with single photons". *European Journal of Physics* **30** (2): 345–353. Bibcode:2009EJPh...30..345B. doi:10.1088/0143-0807/30/2/014.

3.20 External links

- The dictionary definition of photon at Wiktionary

- Media related to Photon at Wikimedia Commons

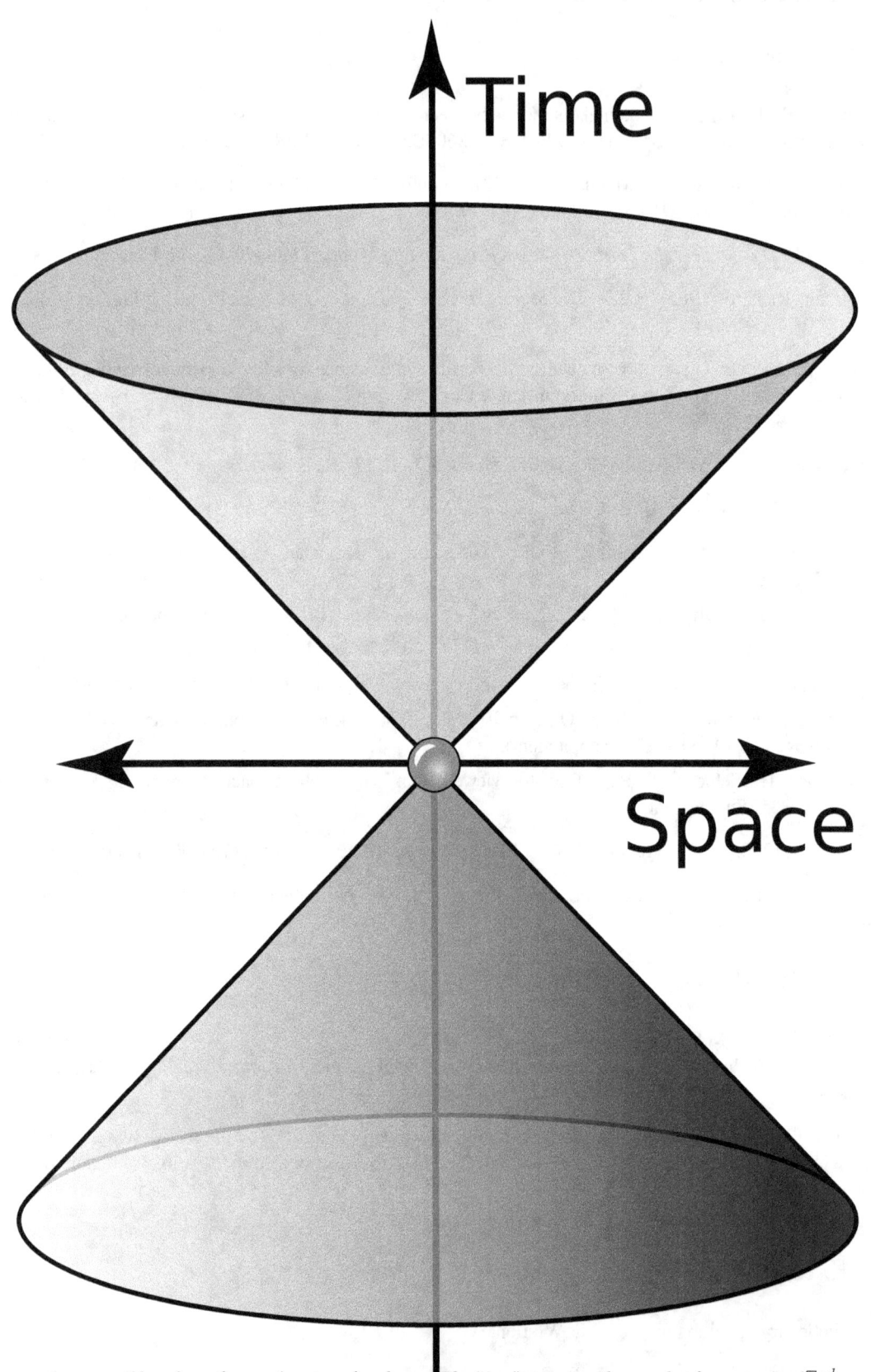

The cone shows possible values of wave 4-vector of a photon. The "time" axis gives the angular frequency (rad⬚s^{-1}

axes represent the angular wavenumber (rad⬚m^{-1}). Green and indigo represent left and right polarization
) and the "space"

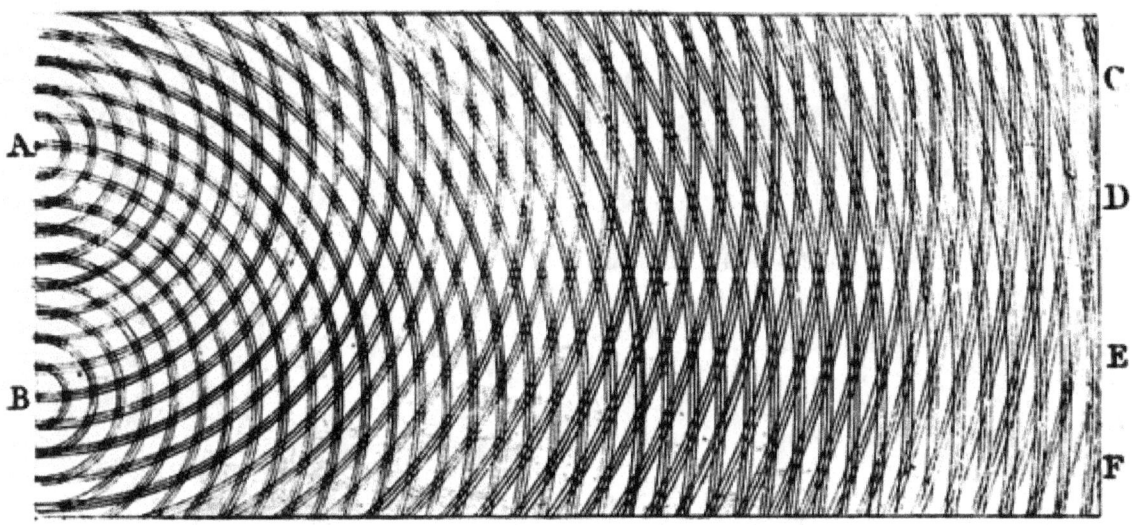

Thomas Young's double-slit experiment in 1801 showed that light can act as a wave, helping to invalidate early particle theories of light.[15]:964

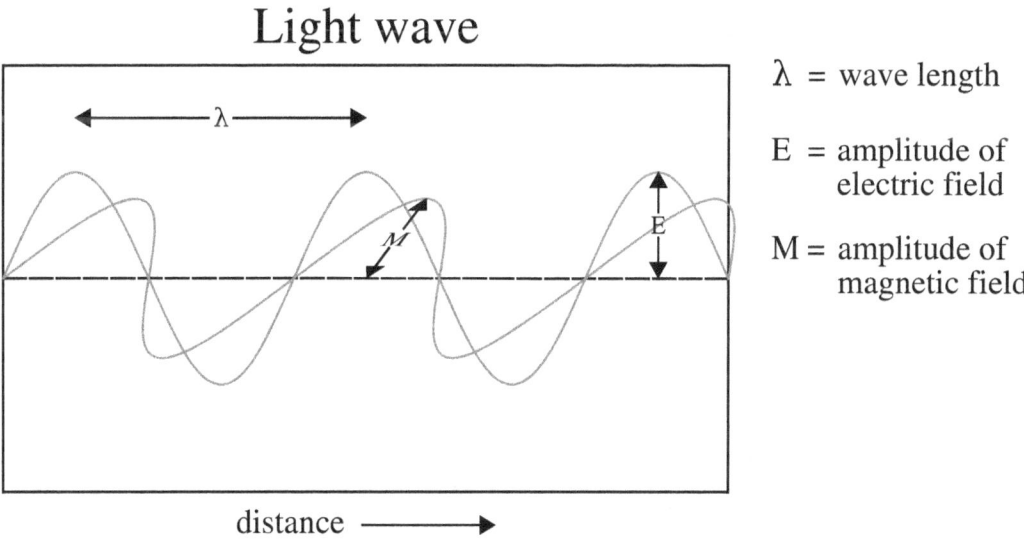

In 1900, Maxwell's theoretical model of light as oscillating electric and magnetic fields seemed complete. However, several observations could not be explained by any wave model of electromagnetic radiation, leading to the idea that light-energy was packaged into quanta described by E=hν. Later experiments showed that these light-quanta also carry momentum and, thus, can be considered particles: the photon concept was born, leading to a deeper understanding of the electric and magnetic fields themselves.

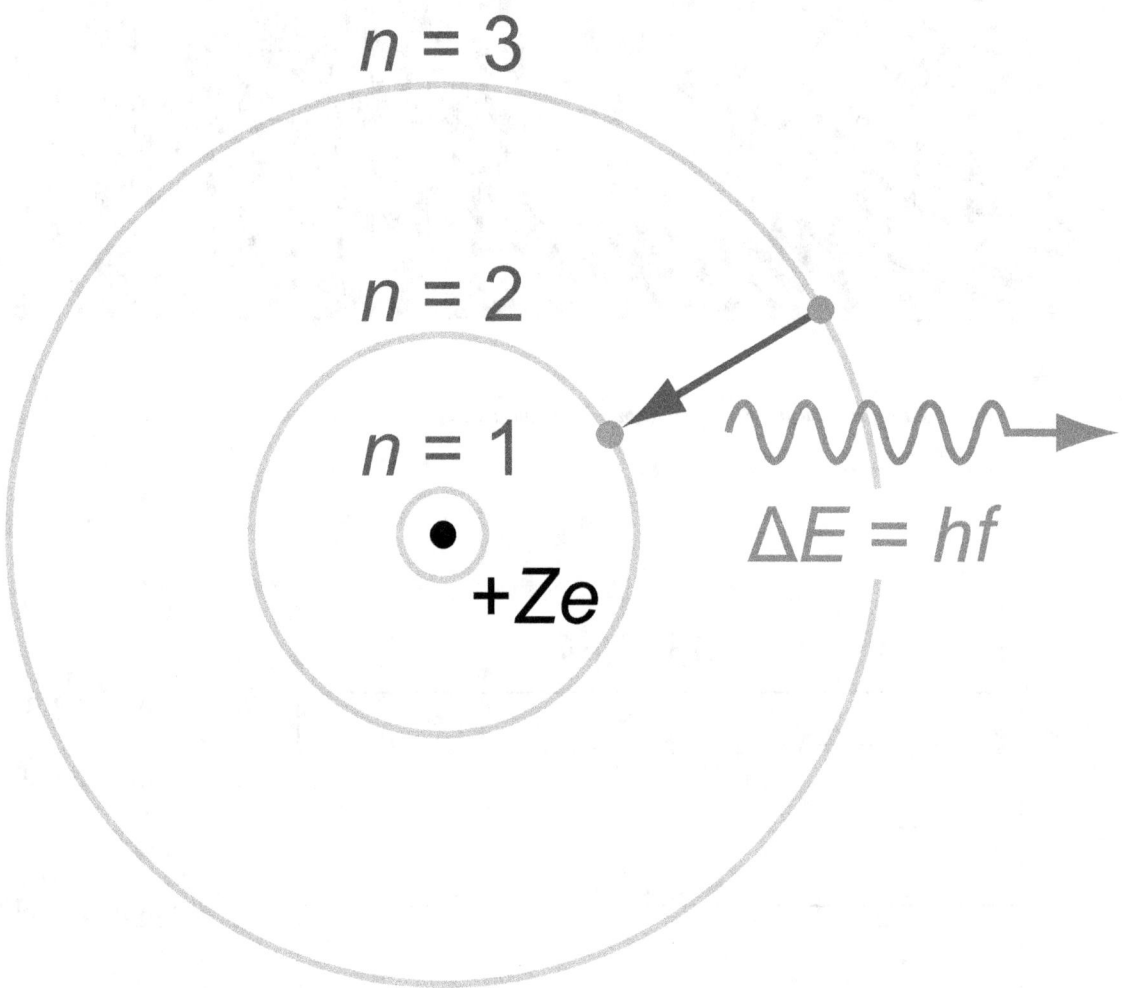

Up to 1923, most physicists were reluctant to accept that light itself was quantized. Instead, they tried to explain photon behavior by quantizing only matter, *as in the Bohr model of the hydrogen atom (shown here). Even though these semiclassical models were only a first approximation, they were accurate for simple systems and they led to quantum mechanics.*

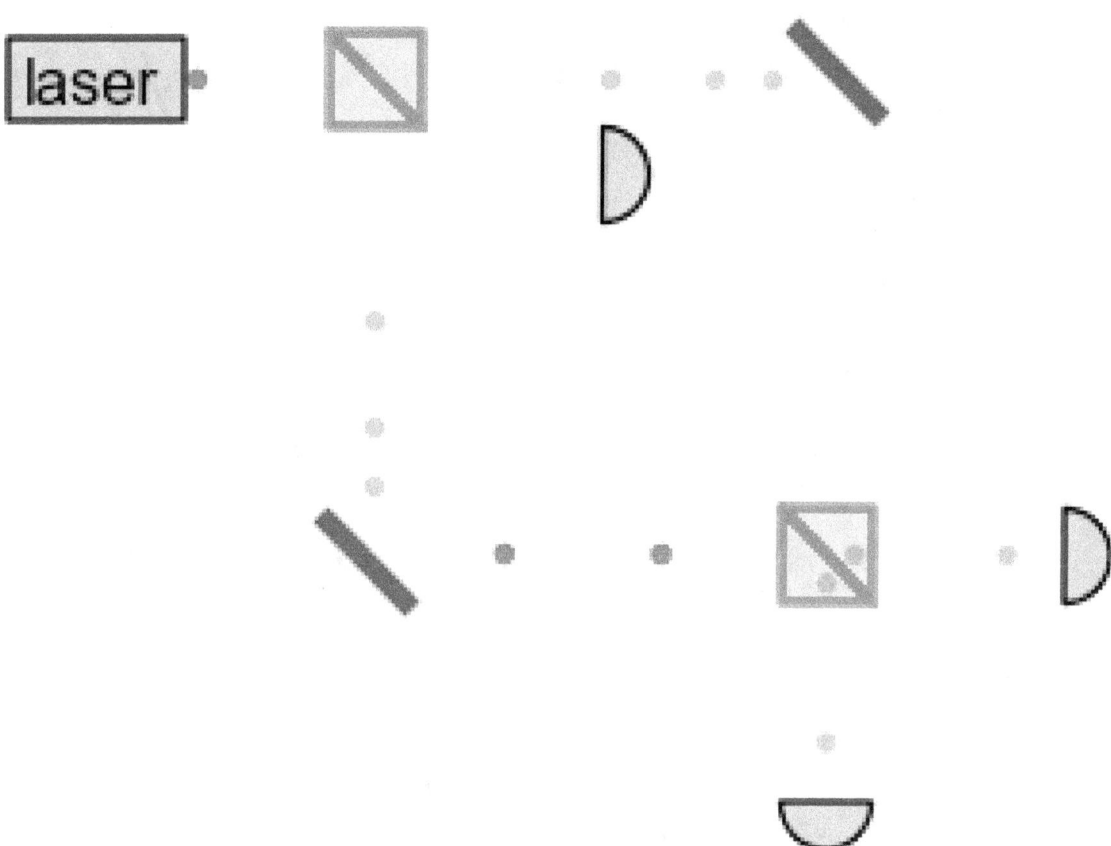

Photons in a Mach–Zehnder interferometer exhibit wave-like interference and particle-like detection at single-photon detectors.

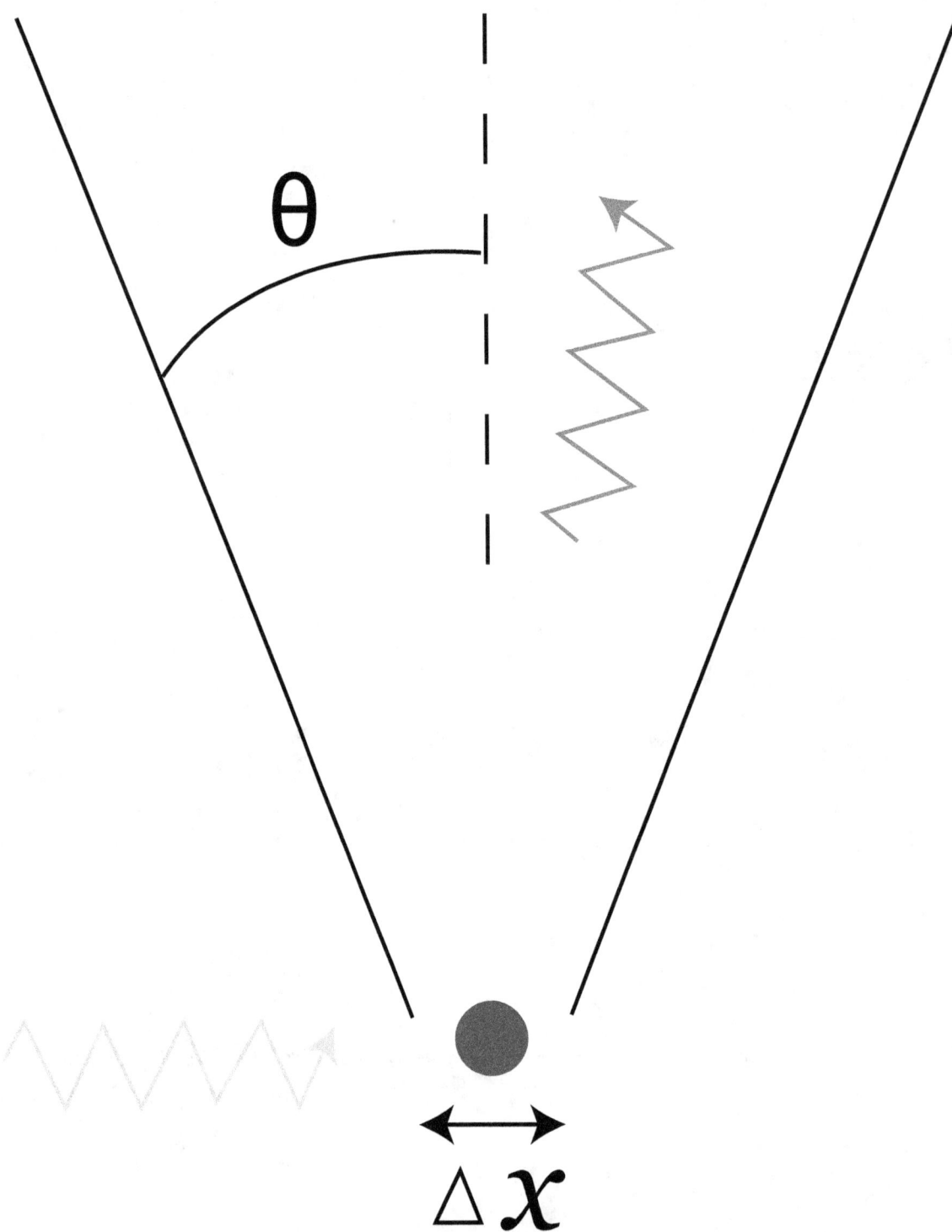

Heisenberg's thought experiment for locating an electron (shown in blue) with a high-resolution gamma-ray microscope. The incoming gamma ray (shown in green) is scattered by the electron up into the microscope's aperture angle θ. The scattered gamma ray is shown in red. Classical optics shows that the electron position can be resolved only up to an uncertainty Δx that depends on θ and the wavelength λ of the incoming light.

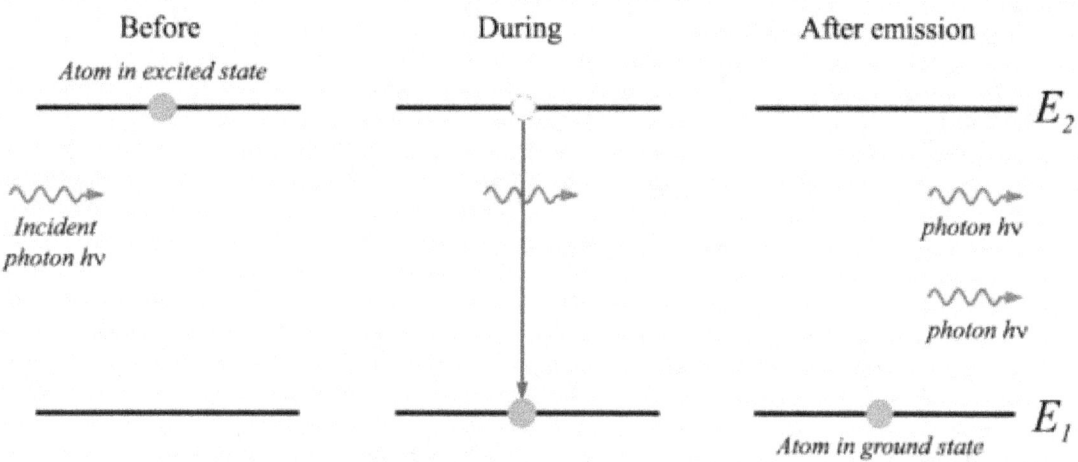

Stimulated emission (in which photons "clone" themselves) was predicted by Einstein in his kinetic analysis, and led to the development of the laser. Einstein's derivation inspired further developments in the quantum treatment of light, which led to the statistical interpretation of quantum mechanics.

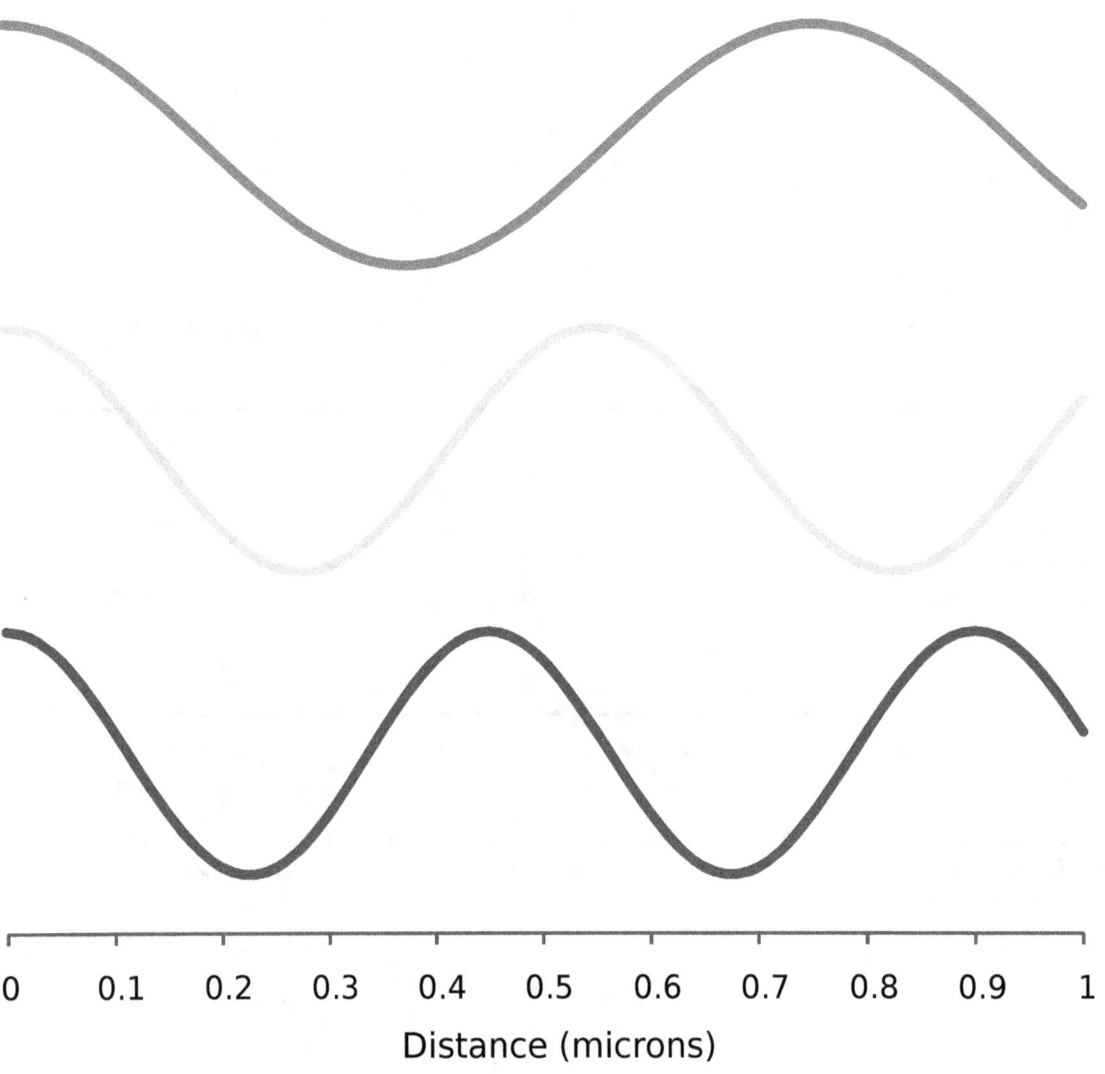

Different electromagnetic modes *(such as those depicted here) can be treated as independent simple harmonic oscillators. A photon corresponds to a unit of energy E=hν in its electromagnetic mode.*

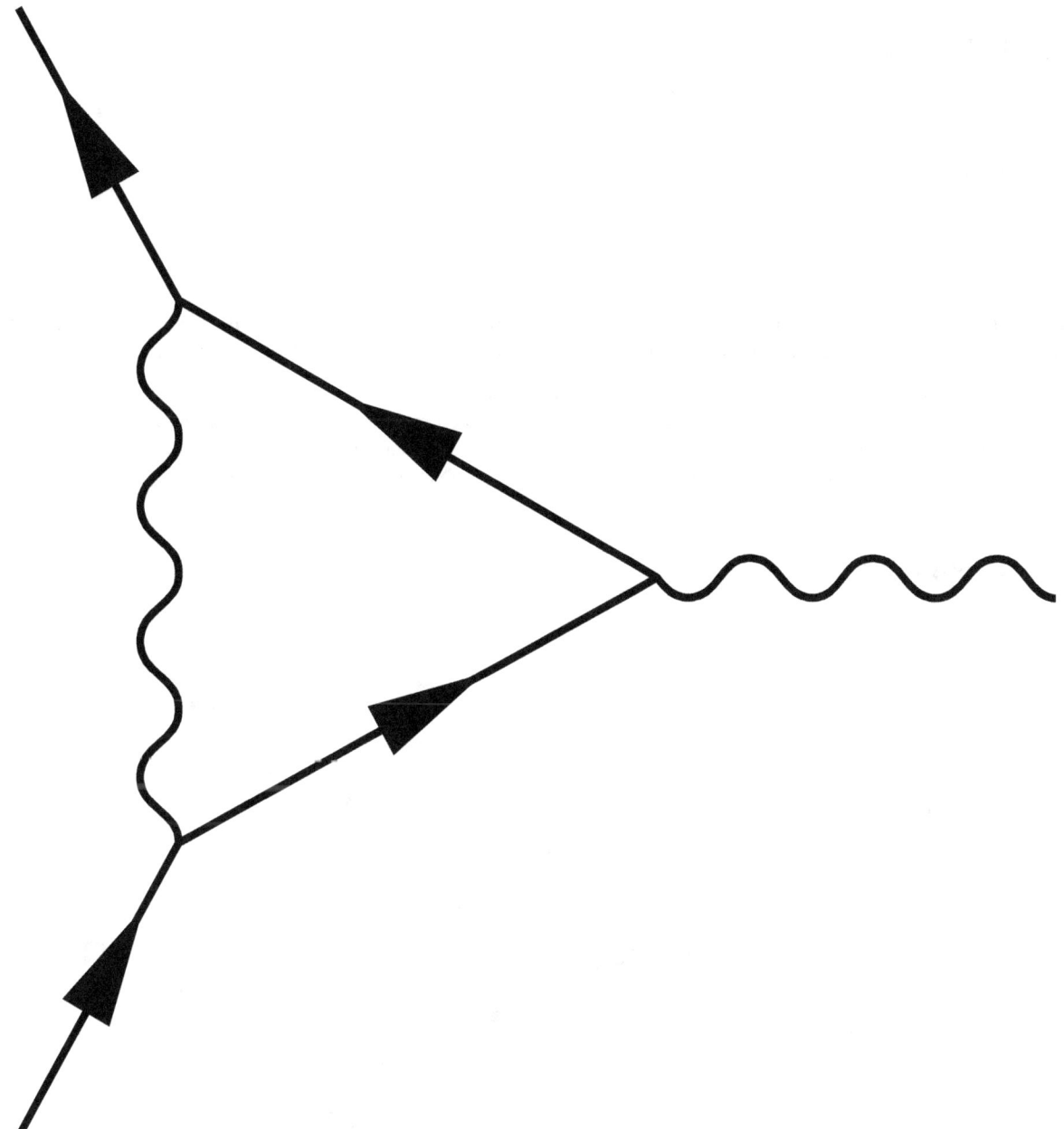

In quantum field theory, the probability of an event is computed by summing the probability amplitude (a complex number) for all possible ways in which the event can occur, as in the Feynman diagram shown here; the probability equals the square of the modulus of the total amplitude.

Chapter 4

Gluon

Gluons /ˈgluːɒnz/ are elementary particles that act as the exchange particles (or gauge bosons) for the strong force between quarks, analogous to the exchange of photons in the electromagnetic force between two charged particles.[6]

In technical terms, gluons are vector gauge bosons that mediate strong interactions of quarks in quantum chromodynamics (QCD). Gluons themselves carry the color charge of the strong interaction. This is unlike the photon, which mediates the electromagnetic interaction but lacks an electric charge. Gluons therefore participate in the strong interaction in addition to mediating it, making QCD significantly harder to analyze than QED (quantum electrodynamics).

4.1 Properties

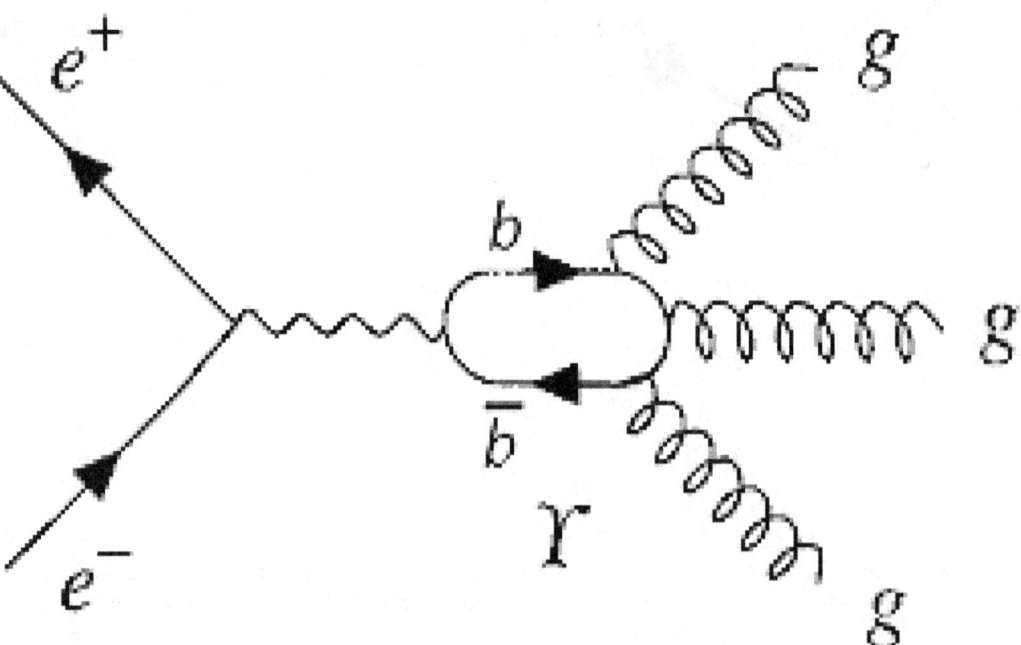

Diagram 2: $e^+e^- \rightarrow \Upsilon(9.46) \rightarrow 3g$

The gluon is a vector boson; like the photon, it has a spin of 1. While massive spin-1 particles have three polarization states,

36

massless gauge bosons like the gluon have only two polarization states because gauge invariance requires the polarization to be transverse. In quantum field theory, unbroken gauge invariance requires that gauge bosons have zero mass (experiment limits the gluon's rest mass to less than a few meV/c^2). The gluon has negative intrinsic parity.

4.2 Numerology of gluons

Unlike the single photon of QED or the three W and Z bosons of the weak interaction, there are eight independent types of gluon in QCD.

This may be difficult to understand intuitively. Quarks carry three types of color charge; antiquarks carry three types of anticolor. Gluons may be thought of as carrying both color and anticolor, but to correctly understand how they are combined, it is necessary to consider the mathematics of color charge in more detail.

4.2.1 Color charge and superposition

In quantum mechanics, the states of particles may be added according to the principle of superposition; that is, they may be in a "combined state" with a *probability*, if some particular quantity is measured, of giving several different outcomes. A relevant illustration in the case at hand would be a gluon with a color state described by:

$(r\bar{b} + b\bar{r})/\sqrt{2}.$

This is read as "red–antiblue plus blue–antired". (The factor of the square root of two is required for normalization, a detail that is not crucial to understand in this discussion.) If one were somehow able to make a direct measurement of the color of a gluon in this state, there would be a 50% chance of it having red-antiblue color charge and a 50% chance of blue-antired color charge.

4.2.2 Color singlet states

It is often said that the stable strongly interacting particles (such as the proton and the neutron, i.e. hadrons) observed in nature are "colorless", but more precisely they are in a "color singlet" state, which is mathematically analogous to a *spin* singlet state.[7] Such states allow interaction with other color singlets, but not with other color states; because long-range gluon interactions do not exist, this illustrates that gluons in the singlet state do not exist either.[7]

The color singlet state is:[7]

$(r\bar{r} + b\bar{b} + g\bar{g})/\sqrt{3}.$

In words, if one could measure the color of the state, there would be equal probabilities of it being red-antired, blue-antiblue, or green-antigreen.

4.2.3 Eight gluon colors

There are eight remaining independent color states, which correspond to the "eight types" or "eight colors" of gluons. Because states can be mixed together as discussed above, there are many ways of presenting these states, which are known as the "color octet". One commonly used list is:[7]

These are equivalent to the Gell-Mann matrices; the translation between the two is that red-antired is the upper-left matrix entry, red-antiblue is the upper middle entry, blue-antigreen is the middle right entry, and so on. The critical feature of these particular eight states is that they are linearly independent, and also independent of the singlet state; there is no way to add any combination of states to produce any other. (It is also impossible to add them to make rr, gg, or bb[8] otherwise the forbidden singlet state could also be made.) There are many other possible choices, but all are mathematically equivalent, at least equally complex, and give the same physical results.

4.2.4   Group theory details

Technically, QCD is a gauge theory with SU(3) gauge symmetry. Quarks are introduced as spinor fields in N_f flavors, each in the fundamental representation (triplet, denoted **3**) of the color gauge group, SU(3). The gluons are vector fields in the adjoint representation (octets, denoted **8**) of color SU(3). For a general gauge group, the number of force-carriers (like photons or gluons) is always equal to the dimension of the adjoint representation. For the simple case of SU(N), the dimension of this representation is $N^2 - 1$.

In terms of group theory, the assertion that there are no color singlet gluons is simply the statement that quantum chromodynamics has an SU(3) rather than a U(3) symmetry. There is no known *a priori* reason for one group to be preferred over the other, but as discussed above, the experimental evidence supports SU(3).[7] The U(1) group for electromagnetic field combines with a slightly more complicated group known as SU(2),S stands for "special", which means the corresponding matrices have derterminant 1.

4.3   Confinement

Main article: Color confinement

Since gluons themselves carry color charge, they participate in strong interactions. These gluon-gluon interactions constrain color fields to string-like objects called "flux tubes", which exert constant force when stretched. Due to this force, quarks are confined within composite particles called hadrons. This effectively limits the range of the strong interaction to 1×10^{-15} meters, roughly the size of an atomic nucleus. Beyond a certain distance, the energy of the flux tube binding two quarks increases linearly. At a large enough distance, it becomes energetically more favorable to pull a quark-antiquark pair out of the vacuum rather than increase the length of the flux tube.

Gluons also share this property of being confined within hadrons. One consequence is that gluons are not directly involved in the nuclear forces between hadrons. The force mediators for these are other hadrons called mesons.

Although in the normal phase of QCD single gluons may not travel freely, it is predicted that there exist hadrons that are formed entirely of gluons — called glueballs. There are also conjectures about other exotic hadrons in which real gluons (as opposed to virtual ones found in ordinary hadrons) would be primary constituents. Beyond the normal phase of QCD (at extreme temperatures and pressures), quark–gluon plasma forms. In such a plasma there are no hadrons; quarks and gluons become free particles.

4.4   Experimental observations

Quarks and gluons (colored) manifest themselves by fragmenting into more quarks and gluons, which in turn hadronize into normal (colorless) particles, correlated in jets. As shown in 1978 summer conferences[2] the PLUTO detector at the electron-positron collider DORIS (DESY) produced the first evidence that the hadronic decays of the very narrow resonance $\Upsilon(9.46)$ could be interpreted as three-jet event topologies produced by three gluons. Later published analyses by the same experiment confirmed this interpretation and also the spin 1 nature of the gluon[9][10] (see also the recollection[2] and PLUTO experiments).

In summer 1979 at higher energies at the electron-positron collider PETRA (DESY) again three-jet topologies were observed, now interpreted as qq gluon bremsstrahlung, now clearly visible, by TASSO,[11] MARK-J[12] and PLUTO experiments[13] (later in 1980 also by JADE[14]). The spin 1 of the gluon was confirmed in 1980 by TASSO[15] and PLUTO experiments[16] (see also the review[3]). In 1991 a subsequent experiment at the LEP storage ring at CERN again confirmed this result.[17]

The gluons play an important role in the elementary strong interactions between quarks and gluons, described by QCD and studied particularly at the electron-proton collider HERA at DESY. The number and momentum distribution of the gluons in the proton (gluon density) have been measured by two experiments, H1 and ZEUS,[18] in the years 1996 till today (2012). The gluon contribution to the proton spin has been studied by the HERMES experiment at HERA.[19] The gluon density in the proton (when behaving hadronically) also has been measured.[20]

Color confinement is verified by the failure of free quark searches (searches of fractional charges). Quarks are normally produced in pairs (quark + antiquark) to compensate the quantum color and flavor numbers; however at Fermilab single production of top quarks has been shown (technically this still involves a pair production, but quark and antiquark are of different flavor).[21] No glueball has been demonstrated.

Deconfinement was claimed in 2000 at CERN SPS[22] in heavy-ion collisions, and it implies a new state of matter: quark–gluon plasma, less interacting than in the nucleus, almost as in a liquid. It was found at the Relativistic Heavy Ion Collider (RHIC) at Brookhaven in the years 2004–2010 by four contemporaneous experiments.[23] A quark–gluon plasma state has been confirmed at the CERN Large Hadron Collider (LHC) by the three experiments ALICE, ATLAS and CMS in 2010.[24]

4.5 See also

- Quark

- Hadron

- Meson

- Gauge boson

- Quark model

- Quantum chromodynamics

- Quark–gluon plasma

- Color confinement

- Glueball

- Gluon field

- Gluon field strength tensor

- Exotic hadrons

- Standard Model

- Three-jet events

- Deep inelastic scattering

4.6 References

[1] M. Gell-Mann (1962). "Symmetries of Baryons and Mesons". *Physical Review* **125** (3): 1067–1084. Bibcode:1962PhRv..125.1067G. doi:10.1103/PhysRev.125.1067.

[2] B.R. Stella and H.-J. Meyer (2011). "ϒ(9.46 GeV) and the gluon discovery (a critical recollection of PLUTO results)". *European Physical Journal H* **36** (2): 203–243. arXiv:1008.1869v3. Bibcode:2011EPJH...36..203S. doi:10.1140/epjh/e2011-10029-3.

[3] P. Söding (2010). "On the discovery of the gluon". *European Physical Journal H* **35** (1): 3–28. Bibcode:2010EPJH...35....3S. doi:10.1140/epjh/e2010-00002-5.

[4] W.-M. Yao; et al. (2006). "Review of Particle Physics" (PDF). *Journal of Physics G* **33**: 1. arXiv:astro-ph/0601168. Bibcode:2006JPhG...33....1Y. doi:10.1088/0954-3899/33/1/001.

[5] F. Yndurain (1995). "Limits on the mass of the gluon". *Physics Letters B* **345** (4): 524. Bibcode:1995PhLB..345..524Y. doi:10.1016/0370-2693(94)01677-5.

[6] C.R. Nave. "The Color Force". *HyperPhysics*. Georgia State University, Department of Physics. Retrieved 2012-04-02.

[7] David Griffiths (1987). *Introduction to Elementary Particles*. John Wiley & Sons. pp. 280–281. ISBN 0-471-60386-4.

[8] J. Baez. "Why are there eight gluons and not nine?". Retrieved 2009-09-13.

[9] Ch. Berger *et al.* (PLUTO Collaboration) (1979). "Jet analysis of the $\Upsilon(9.46)$ decay into charged hadrons". *Physics Letters B* **82** (3–4): 449. Bibcode:1979PhLB...82..449B. doi:10.1016/0370-2693(79)90265-X.

[10] Ch. Berger*et al.* (PLUTO Collaboration) (1981). "Topology of theΥdecay".*Zeitschrift für Physik C***8**(2): 101. Bibcode:1981ZB. doi:10.1007/BF01547873.

[11] R. Brandelik *et al.* (TASSO collaboration) (1979). "Evidence for Planar Events in e^+e^- Annihilation at High Energies". *Physics Letters B* **86** (2): 243–249. Bibcode:1979PhLB...86..243B. doi:10.1016/0370-2693(79)90830-X.

[12] D.P. Barber *et al.* (MARK-J collaboration) (1979). "Discovery of Three-Jet Events and a Test of Quantum Chromodynamics at PETRA". *Physical Review Letters* **43** (12): 830. Bibcode:1979PhRvL..43..830B. doi:10.1103/PhysRevLett.43.830.

[13] Ch. Berger *et al.* (PLUTO Collaboration) (1979). "Evidence for Gluon Bremsstrahlung in e^+e^- Annihilations at High Energies". *Physics Letters B* **86** (3–4): 418. Bibcode:1979PhLB...86..418B. doi:10.1016/0370-2693(79)90869-4.

[14] W. Bartel *et al.* (JADE Collaboration) (1980). "Observation of planar three-jet events in e^+e^- annihilation and evidence for gluon bremsstrahlung". *Physics Letters B* **91**: 142. Bibcode:1980PhLB...91..142B. doi:10.1016/0370-2693(80)90680-2.

[15] R. Brandelik *et al.* (TASSO Collaboration) (1980). "Evidence for a spin-1 gluon in three-jet events". *Physics Letters B* **97** (3–4): 453. Bibcode:1980PhLB...97..453B. doi:10.1016/0370-2693(80)90639-5.

[16] Ch. Berger *et al.* (PLUTO Collaboration) (1980). "A study of multi-jet events in e^+e^- annihilation". *Physics Letters B* **97** (3–4): 459. Bibcode:1980PhLB...97..459B. doi:10.1016/0370-2693(80)90640-1.

[17] G. Alexander *et al.* (OPAL Collaboration) (1991). "Measurement of Three-Jet Distributions Sensitive to the Gluon Spin in e^+e^- Annihilations at $\sqrt{s} = 91$ GeV". *Zeitschrift für Physik C* **52** (4): 543. Bibcode:1991ZPhyC..52..543A. doi:10.1007/BF01562326.

[18] L. Lindeman (H1 and ZEUS collaborations) (1997). "Proton structure functions and gluon density at HERA". *Nuclear Physics B Proceedings Supplements* **64**: 179–183. Bibcode:1998NuPhS..64..179L. doi:10.1016/S0920-5632(97)01057-8.

[19] http://www-hermes.desy.de

[20] C. Adloff *et al.* (H1 collaboration) (1999). "Charged particle cross sections in the photoproduction and extraction of the gluon density in the photon". *European Physical Journal C* **10**: 363–372. arXiv:hep-ex/9810020. Bibcode:1999EPJC...10..363H. doi:10.1007/s100520050761.

[21] M. Chalmers (6 March 2009). "Top result for Tevatron". *Physics World*. Retrieved 2012-04-02.

[22] M.C. Abreu; et al. (2000). "Evidence for deconfinement of quark and antiquark from the J/Ψ suppression pattern measured in Pb-Pb collisions at the CERN SpS". *Physics Letters B* **477**: 28–36. Bibcode:2000PhLB..477...28A. doi:10.1016/S0370-2693(00)00237-9.

[23] D. Overbye (15 February 2010). "In Brookhaven Collider, Scientists Briefly Break a Law of Nature". *New York Times*. Retrieved 2012-04-02.

[24] "LHC experiments bring new insight into primordial universe" (Press release). CERN. 26 November 2010. Retrieved 2012-04-02.

4.7   Further reading

- A. Ali and G. Kramer (2011). "JETS and QCD: A historical review of the discovery of the quark and gluon jets and its impact on QCD". *European Physical Journal H* **36** (2): 245–326. arXiv:1012.2288. Bibcode:2011EPJH...36. A.doi:10.1140/epjh/e2011-10047-1.

Chapter 5

W and Z bosons

The **W and Z bosons** (together known as the **weak bosons** or, less specifically, the **intermediate vector bosons**) are the elementary particles that mediate the weak interaction; their symbols are W+, W−, and Z. The W bosons have a positive and negative electric charge of 1 elementary charge respectively and are each other's antiparticles. The Z boson is electrically neutral and is its own antiparticle. The three particles have a spin of 1, and the W bosons have a magnetic moment, while the Z has none. All three of these particles are very short-lived, with a half-life of about $3{\times}10^{-25}$ s. Their discovery was a major success for what is now called the Standard Model of particle physics.

The W bosons are named after the weak force. The physicist Steven Weinberg named the additional particle the "Z particle",[3] later giving the explanation that it was the last additional particle needed by the model – the W bosons had already been named – and that it has zero electric charge.[4]

The two **W bosons** are best known as mediators of neutrino absorption and emission, where their charge is associated with electron or positron emission or absorption, always causing nuclear transmutation. The Z boson is not involved in the absorption or emission of electrons and positrons.

The **Z boson** mediates the transfer of momentum, spin, and energy when neutrinos scatter *elastically* from matter, something that must happen without the production or absorption of new, charged particles. Such behaviour (which is almost as common as inelastic neutrino interactions) is seen in bubble chambers irradiated with neutrino beams. Whenever an electron simply "appears" in such a chamber as a new free particle suddenly moving with kinetic energy, and moves in the direction of the neutrinos as the apparent result of a new impulse, and this behavior happens more often when the neutrino beam is present, it is inferred to be a result of a neutrino interacting directly with the electron. Here, the neutrino simply strikes the electron and scatters away from it, transferring some of the neutrino's momentum to the electron. Since (i) neither neutrinos nor electrons are affected by the strong force, (ii) neutrinos are electrically neutral (therefore don't interact electromagnetically), and (iii) the incredibly small masses of these particles make any gravitational force between them negligible, such an interaction can only happen via the weak force. Since such an electron is not created from a nucleon, and is unchanged except for the new force impulse imparted by the neutrino, this weak force interaction between the neutrino and the electron must be mediated by a weak-force boson particle with no charge. Thus, this interaction requires a Z boson.

5.1 Basic properties

These bosons are among the heavyweights of the elementary particles. With masses of 80.4 GeV/c^2 and 91.2 GeV/c^2, respectively, the W and Z bosons are almost 100 times as large as the proton – heavier, even, than entire atoms of iron. The masses of these bosons are significant because they act as the force carriers of a quite short-range fundamental force: their high masses thus limit the range of the weak nuclear force. By way of contrast, the electromagnetic force has an infinite range, because its force carrier, the photon, has zero mass, and the same is supposed of the hypothetical graviton.

All three bosons have particle spin $s = 1$. The emission of a W+ or W− boson either raises or lowers the electric charge of the emitting particle by one unit, and also alters the spin by one unit. At the same time, the emission or absorption of a W boson can change the type of the particle – for example changing a strange quark into an up quark. The neutral Z boson cannot change the electric charge of any particle, nor can it change any other of the so-called "charges" (such as strangeness, baryon number, charm, etc.). The emission or absorption of a Z boson can only change the spin, momentum, and energy of the other particle. (See also *weak neutral current*.)

5.2   Weak nuclear force

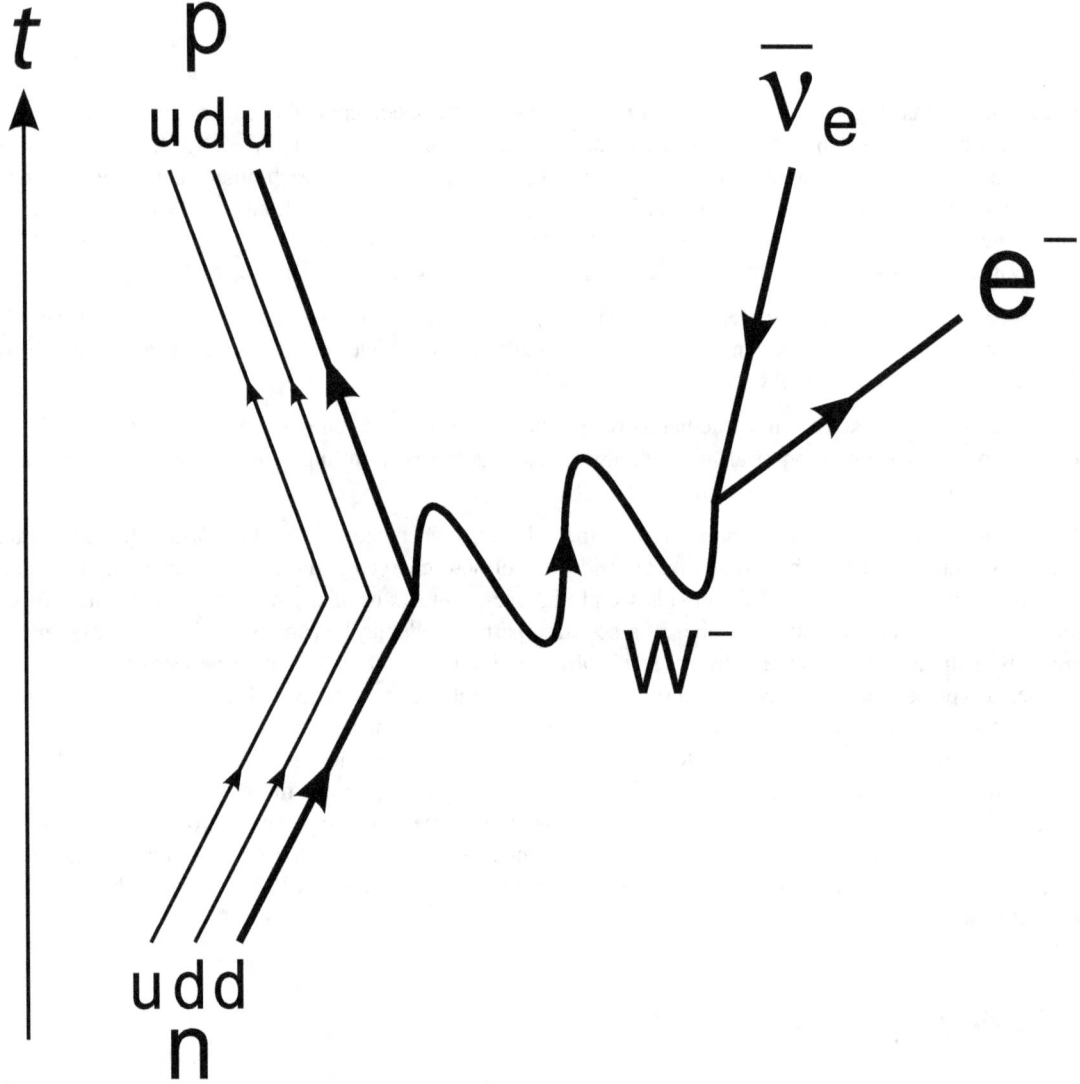

The Feynman diagram for beta decay of a neutron into a proton, electron, and electron antineutrino via an intermediate heavy W boson

The W and Z bosons are carrier particles that mediate the weak nuclear force, much as the photon is the carrier particle for the electromagnetic force.

5.2.1 W bosons

The W bosons are best known for their role in nuclear decay. Consider, for example, the beta decay of cobalt-60.

$$^{60}_{27}\text{Co} \rightarrow ^{60}_{28}\text{Ni}^+ + e- + \nu_e$$

This reaction does not involve the whole cobalt-60 nucleus, but affects only one of its 33 neutrons. The neutron is converted into a proton while also emitting an electron (called a beta particle in this context) and an electron antineutrino:

$$n0 \rightarrow p+ + e- + \nu_e$$

Again, the neutron is not an elementary particle but a composite of an up quark and two down quarks (udd). It is in fact one of the down quarks that interacts in beta decay, turning into an up quark to form a proton (uud). At the most fundamental level, then, the weak force changes the flavour of a single quark:

$$d \rightarrow u + W-$$

which is immediately followed by decay of the W− itself:

$$W- \rightarrow e- + \nu_e$$

5.2.2 Z boson

The Z boson is its own antiparticle. Thus, all of its flavour quantum numbers and charges are zero. The exchange of a Z boson between particles, called a neutral current interaction, therefore leaves the interacting particles unaffected, except for a transfer of momentum. Z boson interactions involving neutrinos have distinctive signatures: They provide the only known mechanism for elastic scattering of neutrinos in matter; neutrinos are almost as likely to scatter elastically (via Z boson exchange) as inelastically (via W boson exchange). The first prediction of Z bosons was made by Brazilian physicist José Leite Lopes in 1958,[5] by devising an equation which showed the analogy of the weak nuclear interactions with electromagnetism. Steve Weinberg, Sheldon Glashow and Abdus Salam used later these results to develop the electroweak unification,[6] in 1973. Weak neutral currents via Z boson exchange were confirmed shortly thereafter in 1974, in a neutrino experiment in the Gargamelle bubble chamber at CERN.

5.3 Predicting the W and Z

Following the spectacular success of quantum electrodynamics in the 1950s, attempts were undertaken to formulate a similar theory of the weak nuclear force. This culminated around 1968 in a unified theory of electromagnetism and weak interactions by Sheldon Glashow, Steven Weinberg, and Abdus Salam, for which they shared the 1979 Nobel Prize in Physics.[7] Their electroweak theory postulated not only the W bosons necessary to explain beta decay, but also a new Z boson that had never been observed.

The fact that the W and Z bosons have mass while photons are massless was a major obstacle in developing electroweak theory. These particles are accurately described by an SU(2) gauge theory, but the bosons in a gauge theory must be massless. As a case in point, the photon is massless because electromagnetism is described by a U(1) gauge theory. Some mechanism is required to break the SU(2) symmetry, giving mass to the W and Z in the process. One explanation, the Higgs mechanism, was forwarded by the 1964 PRL symmetry breaking papers. It predicts the existence of yet another

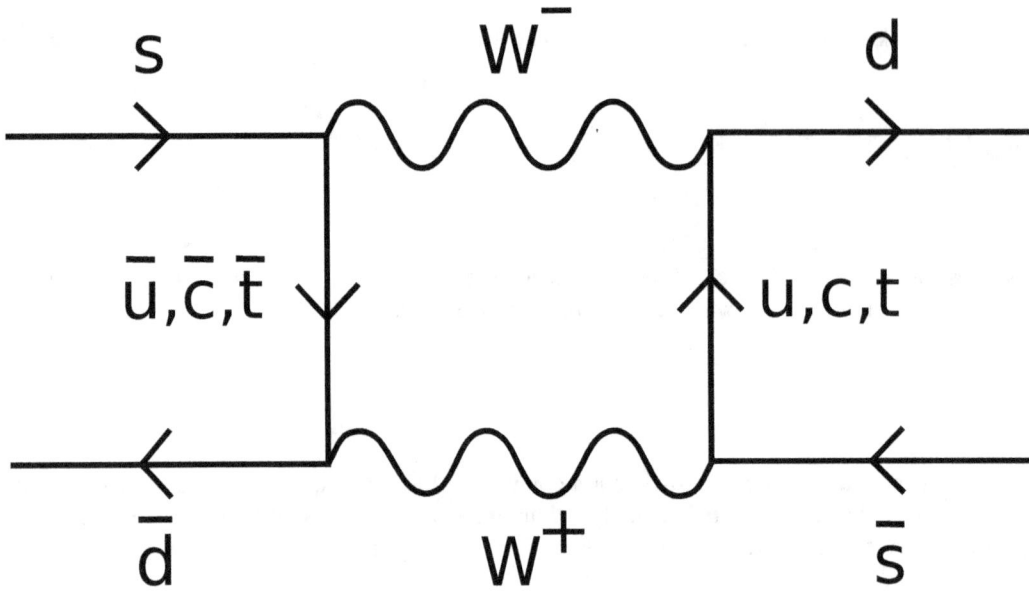

A Feynman diagram showing the exchange of a pair of W bosons. This is one of the leading terms contributing to neutral Kaon oscillation.

new particle; the Higgs boson. Of the four components of a Goldstone boson created by the Higgs field, three are "eaten" by the W^+, Z^0, and W^- bosons to form their longitudinal components and the remainder appears as the spin 0 Higgs boson.

The combination of the SU(2) gauge theory of the weak interaction, the electromagnetic interaction, and the Higgs mechanism is known as the Glashow-Weinberg-Salam model. These days it is widely accepted as one of the pillars of the Standard Model of particle physics. As of 13 December 2011, intensive search for the Higgs boson carried out at CERN has indicated that if the particle is to be found, it seems likely to be found around 125 GeV. On 4 July 2012, the CMS and the ATLAS experimental collaborations at CERN announced the discovery of a new particle with a mass of 125.3 ± 0.6 GeV that appears consistent with a Higgs boson.

5.4   Discovery

Unlike beta decay, the observation of neutral current interactions that involve particles *other than neutrinos* requires huge investments in particle accelerators and detectors, such as are available in only a few high-energy physics laboratories in the world (and then only after 1983). This is because Z-bosons behave in somewhat the same manner as photons, but do not become important until the energy of the interaction is comparable with the relatively huge mass of the Z boson.

The discovery of the W and Z bosons was considered a major success for CERN. First, in 1973, came the observation of neutral current interactions as predicted by electroweak theory. The huge Gargamelle bubble chamber photographed the tracks of a few electrons suddenly starting to move, seemingly of their own accord. This is interpreted as a neutrino interacting with the electron by the exchange of an unseen Z boson. The neutrino is otherwise undetectable, so the only observable effect is the momentum imparted to the electron by the interaction.

The discovery of the W and Z bosons themselves had to wait for the construction of a particle accelerator powerful enough to produce them. The first such machine that became available was the Super Proton Synchrotron, where unambiguous signals of W bosons were seen in January 1983 during a series of experiments made possible by Carlo Rubbia and Simon van der Meer. The actual experiments were called UA1 (led by Rubbia) and UA2 (led by Pierre Darriulat),[8] and were the collaborative effort of many people. Van der Meer was the driving force on the accelerator end (stochastic cooling). UA1 and UA2 found the Z boson a few months later, in May 1983. Rubbia and van der Meer were promptly awarded

The Gargamelle bubble chamber, now exhibited at CERN

the 1984 Nobel Prize in Physics, a most unusual step for the conservative Nobel Foundation.[9]

The W+, W−, and Z0 bosons, together with the photon (γ), comprise the four gauge bosons of the electroweak interaction.

5.5 Decay

The W and Z bosons decay to fermion–antifermion pairs but neither the W nor the Z bosons can decay into the higher-mass top quark. Neglecting phase space effects and higher order corrections, simple estimates of their branching fractions can be calculated from the coupling constants.

5.5.1 W bosons

W bosons can decay to a lepton and neutrino or to an up-type quark and a down-type quark. The decay width of the W boson to a quark–antiquark pair is proportional to the corresponding squared CKM matrix element and the number of quark colours, $NC = 3$. The decay widths for the W bosons are then proportional to:

Here, e+, μ+, τ+ denote the three flavours of leptons (more exactly, the positive charged antileptons). ν
e, ν
μ, ν
τ denote the three flavours of neutrinos. The other particles, starting with u and d, all denote quarks and antiquarks (factor NC is applied). The various Vij denote the corresponding CKM matrix coefficients.

Unitarity of the CKM matrix implies that $|V_{ud}|^2 + |V_{us}|^2 + |V_{ub}|^2 = |V_{cd}|^2 + |V_{cs}|^2 + |V_{cb}|^2 = 1$. Therefore the leptonic branching ratios of the W boson are approximately B(e+ν
e) = B(μ+ν

μ) = $B(\tau + \nu$

τ) = $\frac{1}{9}$. The hadronic branching ratio is dominated by the CKM-favored ud and cs final states. The sum of the hadronic branching ratios has been measured experimentally to be 67.60±0.27%, with $B(l^+\nu_l) = 10.80\pm0.09\%$.[1]

5.5.2   Z bosons

Z bosons decay into a fermion and its antiparticle. As the Z-boson is a mixture of the pre-symmetry-breaking W^0 and B^0 bosons (see weak mixing angle), each vertex factor includes a factor $T_3 - Q\sin^2\theta W$, where T_3 is the third component of the weak isospin of the fermion, Q is the electric charge of the fermion (in units of the elementary charge), and θW is the weak mixing angle. Because the weak isospin is different for fermions of different chirality, either left-handed or right-handed, the coupling is different as well.

The **relative** strengths of each coupling can be estimated by considering that the decay rates include the square of these factors, and all possible diagrams (e.g. sum over quark families, and left and right contributions). This is just an estimate, as we are considering only tree-level diagrams in the Fermi theory.

Here, L and R denote the left- and right-handed chiralities of the fermions respectively. (The right-handed neutrinos do not exist in the standard model. However, in some extensions beyond the standard model they do.) The notation $x = \sin^2\theta W$ is used.

5.6   See also

- Bose–Einstein statistics

- Boson

- List of particles

- Standard Model (mathematical formulation)

- W' and Z' bosons

- X and Y bosons: analogous pair of bosons predicted by the Grand Unified Theory

5.7   References

[1] J. Beringer; et al. (2012). "2012 Review of Particle Physics - Gauge and Higgs Bosons" (PDF). *Physical Review D* **86**: 1. Bibcode:2012PhRvD..86a0001B. doi:10.1103/PhysRevD.86.010001.

[2] (PDF) http://pdg.lbl.gov/2013/reviews/rpp2013-rev-w-mass.pdf. Missing or empty |title= (help)

[3] Steven Weinberg, A Model of Leptons, Phys. Rev. Lett. 19, 1264–1266 (1967) – the electroweak unification paper.

[4] Weinberg, Steven (1993). *Dreams of a Final Theory: the search for the fundamental laws of nature.* Vintage Press. p. 94. ISBN 0-09-922391-0.

[5] "Forty years of the first attempt at the electroweak unification and of the prediction of the weak neutral boson".

[6] "The Nobel Prize in Physics 1979". Nobel Foundation. Retrieved 2008-09-10.

[7] Nobel Prize in Physics for 1979 (see also Nobel Prize in Physics on Wikipedia)

[8] The UA2 Collaboration collection

[9] 1984 Nobel Prize in physics

[10] C. Amsler et al. (Particle Data Group), PL B667, 1 (2008) and 2009 partial update for the 2010 edition

5.8 External links

- The Review of Particle Physics, the ultimate source of information on particle properties.

- The W and Z particles: a personal recollection by Pierre Darriulat

- When CERN saw the end of the alphabet by Daniel Denegri

- W and Z particles at Hyperphysics

Chapter 6

Scalar boson

A **scalar boson** is a boson whose spin equals zero. *Boson* means that it has an integer-valued spin; the *scalar* fixes this value to 0.

The name "scalar boson" arises from quantum field theory. It refers to the particular transformation properties under Lorentz transformation.

6.1 Examples

- Various known composite particles are scalar bosons, e.g. the alpha particle and the pi meson. Among the scalar mesons, one distinguishes between the scalar and pseudoscalar mesons, which refers to their transformation property under parity.

- The only fundamental scalar boson in the standard model of elementary particle physics is the Higgs boson, whose existence was confirmed on 14 March 2013 at the Large Hadron Collider. As a result of this confirmation, the 2013 Nobel Prize in physics was awarded to Peter Higgs and François Englert.

- One very popular quantum field theory, which uses scalar bosonic fields and is introduced in many introductory books to quantum field theories[1] for pedagogical reasons, is the so-called φ^4-theory. It usually serves as a toy model to introduce the basic concepts of the field.

6.2 See also

- Scalar meson
- Scalar field theory
- Vector boson
- Higgs Boson

6.3 References

[1] Michael E. Peskin and Daniel V. Schroeder (1995). *An Introduction to Quantum Field Theory*. Westview Press. ISBN 0-201-50397-2.

Chapter 7

Higgs boson

The **Higgs boson** or **Higgs particle** is an elementary particle in the Standard Model of particle physics. It is the quantum excitation of the **Higgs field**[6][7]—a fundamental field of crucial importance to particle physics theory,[7] first suspected to exist in the 1960s, that unlike other known fields such as the electromagnetic field, takes a non-zero constant value almost everywhere. The question of the Higgs field's existence has been the last unverified part of the Standard Model of particle physics and, according to some, "the central problem in particle physics".[8][9] The presence of this field, now believed to be confirmed, explains why some fundamental particles have mass when, based on the symmetries controlling their interactions, they should be massless. The existence of the Higgs field would also resolve several other long-standing puzzles, such as the reason for the weak force's extremely short range.

Although it is hypothesized that the Higgs field permeates the entire Universe, evidence for its existence has been very difficult to obtain. In principle, the Higgs field can be detected through its excitations, manifest as Higgs particles, but these are extremely difficult to produce and detect. The importance of this fundamental question led to a 40 year search, and the construction of one of the world's most expensive and complex experimental facilities to date, CERN's Large Hadron Collider,[10] able to create Higgs bosons and other particles for observation and study. On 4 July 2012, the discovery of a new particle with a mass between 125 and 127 GeV/c^2 was announced; physicists suspected that it was the Higgs boson.[11][12][13] Since then, however, the particle had been shown to behave, interact, and decay in many of the ways predicted by the Standard Model. It was also tentatively confirmed to have even parity and zero spin,[1] two fundamental attributes of a Higgs boson. This appears to be the first elementary scalar particle discovered in nature.[14] More data are needed to verify that the discovered particle has properties matching those predicted for the Higgs boson by the Standard Model, or whether, as predicted by some theories, multiple Higgs bosons exist.[3]

The Higgs boson is named after Peter Higgs, one of six physicists who, in 1964, proposed the mechanism that suggested the existence of such a particle. On December 10, 2013, two of them, Peter Higgs and François Englert, were awarded the Nobel Prize in Physics for their work and prediction (Englert's co-researcher Robert Brout had died in 2011 and the Nobel Prize is not ordinarily given posthumously).[15] Although Higgs's name has come to be associated with this theory, several researchers between about 1960 and 1972 each independently developed different parts of it. In mainstream media the Higgs boson has often been called the "God particle", from a 1993 book on the topic; the nickname is strongly disliked by many physicists, including Higgs, who regard it as sensationalistic.[16][17][18]

In the Standard Model, the Higgs particle is a boson with no spin, electric charge, or colour charge. It is also very unstable, decaying into other particles almost immediately. It is a quantum excitation of one of the four components of the Higgs field. The latter constitutes a scalar field, with two neutral and two electrically charged components that form a complex doublet of the weak isospin SU(2) symmetry. The Higgs field is tachyonic (this does not refer to faster-than-light speeds, it means that symmetry-breaking through condensation of a particle must occur under certain conditions), and has a "Mexican hat" shaped potential with nonzero strength everywhere (including otherwise empty space), which in its vacuum state breaks the weak isospin symmetry of the electroweak interaction. When this happens, three components of the Higgs field are "absorbed" by the SU(2) and U(1) gauge bosons (the "Higgs mechanism") to become the longitudinal components of the now-massive W and Z bosons of the weak force. The remaining electrically neutral component separately couples to other particles known as fermions (via Yukawa couplings), causing these to acquire mass as well. Some versions of the theory predict more than one kind of Higgs fields and bosons. Alternative "Higgsless" models would have been considered if the Higgs boson was not discovered.

7.1   A non-technical summary

7.1.1   "Higgs" terminology

7.1.2   Overview

Physicists explain the properties and forces between elementary particles in terms of the Standard Model—a widely accepted and "remarkably" accurate[21] framework based on gauge invariance and symmetries, believed to explain almost everything in the known universe, other than gravity.[22] But by around 1960 all attempts to create a gauge invariant theory for two of the four fundamental forces had consistently failed at one crucial point: although gauge invariance seemed extremely important, it seemed to make any theory of electromagnetism and the weak force go haywire, by demanding that either many particles with mass were massless or that non-existent forces and massless particles had to exist. Scientists had no idea how to get past this point.

In 1962 physicist Philip Anderson wrote a paper that built upon work by Yoichiro Nambu concerning "broken symmetries" in superconductivity and particle physics. He suggested that "broken symmetries" might also be the missing piece needed to solve the problems of gauge invariance. In 1964 a theory was created almost simultaneously by 3 different groups of researchers, that showed Anderson's suggestion was possible - the gauge theory and "mass problems" could indeed be resolved if an unusual kind of field, now generally called the "Higgs field", existed throughout the universe; if the Higgs field did exist, it would apparently cause existing particles to acquire mass instead of new massless particles being formed. Although these ideas did not gain much initial support or attention, by 1972 they had been developed into a comprehensive theory and proved capable of giving "sensible" results that accurately described particles known at the time, and which accurately predicted of several other particles discovered during the following years.[Note 7] During the 1970s these theories rapidly became the "standard model". There was not yet any direct evidence that the Higgs field actually existed, but even without proof of the field, the accuracy of its predictions led scientists to believe the theory might be true. By the 1980s the question whether or not the Higgs field existed had come to be regarded as one of the most important unanswered questions in particle physics.

If Higgs field could be shown to exist, it would be a monumental discovery for science and human knowledge, and would open doorways to new knowledge in many disciplines. If not, then other more complicated theories would need to be considered. The simplest means to test the existence of the Higgs field would be a search for a new elementary particle that the field would have to give off, a particle known as "Higgs bosons" or the "Higgs particle". This particle would be extremely difficult to find. After significant technological advancements, by the 1990s two large experimental installations were being designed and constructed that allowed to search for the Higgs boson.

While several symmetries in nature are spontaneously broken through a form of the Higgs mechanism, in the context of the Standard Model the term "Higgs mechanism" almost always means symmetry breaking of the electroweak field. It is considered confirmed, but revealing the exact cause has been difficult. Various analogies have also been invented to describe the Higgs field and boson, including analogies with well-known symmetry breaking effects such as the rainbow and prism, electric fields, ripples, and resistance of macro objects moving through media, like people moving through crowds or some objects moving through syrup or molasses. However, analogies based on simple resistance to motion are inaccurate as the Higgs field does not work by resisting motion.

7.2   Significance

7.2.1 Scientific impact

Evidence of the Higgs field and its properties has been extremely significant scientifically, for many reasons. The Higgs boson's importance is largely that it is able to be examined using existing knowledge and experimental technology, as a way to confirm and study the entire Higgs field theory.[6][7] Conversely, proof that the Higgs field and boson do not exist would also have been significant. In discussion form, the relevance includes:

7.2.2 Practical and technological impact of discovery

As yet, there are no known immediate technological benefits of finding the Higgs particle. However, a common pattern for fundamental discoveries is for practical applications to follow later, once the discovery has been explored further, at which point they become the basis for new technologies of importance to society.[44][45][46]

The challenges in particle physics have furthered major technological of widespread importance. For example, the World Wide Web began as a project to improve CERN's communication system. CERN's requirement to process massive amounts of data produced by the Large Hadron Collider also led to contributions to the fields of distributed and cloud computing.

7.3 History

See also: 1964 PRL symmetry breaking papers, Higgs mechanism and History of quantum field theory
 Particle physicists study matter made from fundamental particles whose interactions are mediated by exchange particles - gauge bosons - acting as force carriers. At the beginning of the 1960s a number of these particles had been discovered or proposed, along with theories suggesting how they relate to each other, some of which had already been reformulated as field theories in which the objects of study are not particles and forces, but quantum fields and their symmetries.[47]:150 However, attempts to unify known fundamental forces such as the electromagnetic force and the weak nuclear force were known to be incomplete. One known omission was that gauge invariant approaches, including non-abelian models such as Yang–Mills theory (1954), which held great promise for unified theories, also seemed to predict known massive particles as massless.[48] Goldstone's theorem, relating to continuous symmetries within some theories, also appeared to rule out many obvious solutions,[49] since it appeared to show that zero-mass particles would have to also exist that were "simply not seen".[50] According to Guralnik, physicists had "no understanding" how these problems could be overcome.[50]

Particle physicist and mathematician Peter Woit summarised the state of research at the time:

> "Yang and Mills work on non-abelian gauge theory had one huge problem: in perturbation theory it has massless particles which don't correspond to anything we see. One way of getting rid of this problem is now fairly well-understood, the phenomenon of confinement realized in QCD, where the strong interactions get rid of the massless "gluon" states at long distances. By the very early sixties, people had begun to understand another source of massless particles: spontaneous symmetry breaking of a continuous symmetry. What Philip Anderson realized and worked out in the summer of 1962 was that, when you have *both* gauge symmetry *and* spontaneous symmetry breaking, the Nambu–Goldstone massless mode can combine with the massless gauge field modes to produce a physical massive vector field. This is what happens in superconductivity, a subject about which Anderson was (and is) one of the leading experts." *[text condensed]* [48]

The Higgs mechanism is a process by which vector bosons can get rest mass *without* explicitly breaking gauge invariance, as a byproduct of spontaneous symmetry breaking.[51][52] The mathematical theory behind spontaneous symmetry breaking was initially conceived and published within particle physics by Yoichiro Nambu in 1960,[53] the concept that such a mechanism could offer a possible solution for the "mass problem" was originally suggested in 1962 by Philip Anderson (who had previously written papers on broken symmetry and its outcomes in superconductivity[54] and concluded in his 1963 paper on Yang-Mills theory that *"considering the superconducting analog... [t]hese two types of bosons seem capable of canceling each other out... leaving finite mass bosons"*),[55]:4–5[56] and Abraham Klein and Benjamin Lee showed in March 1964 that Goldstone's theorem could be avoided this way in at least some non-relativistic cases and speculated it might be possible in truly relativistic cases.[57]

Nobel Prize Laureate Peter Higgs in Stockholm, December 2013

These approaches were quickly developed into a full relativistic model, independently and almost simultaneously, by three groups of physicists: by François Englert and Robert Brout in August 1964;[58] by Peter Higgs in October 1964;[59] and by Gerald Guralnik, Carl Hagen, and Tom Kibble (GHK) in November 1964.[60] Higgs also wrote a short but important[51] response published in September 1964 to an objection by Gilbert,[61] which showed that if calculating within the radiation gauge, Goldstone's theorem and Gilbert's objection would become inapplicable.[Note 11] (Higgs later described Gilbert's objection as prompting his own paper.[62]) Properties of the model were further considered by Guralnik in 1965,[63] by Higgs in 1966,[64] by Kibble in 1967,[65] and further by GHK in 1967.[66] The original three 1964 papers showed that when a gauge theory is combined with an additional field that spontaneously breaks the symmetry, the gauge bosons can consistently acquire a finite mass.[51][52][67] In 1967, Steven Weinberg[68] and Abdus Salam[69] independently showed how a Higgs mechanism could be used to break the electroweak symmetry of Sheldon Glashow's unified model for the weak and electromagnetic interactions[70] (itself an extension of work by Schwinger), forming what became the Standard Model of particle physics. Weinberg was the first to observe that this would also provide mass terms for the fermions.[71] [Note 12]

However, the seminal papers on spontaneous breaking of gauge symmetries were at first largely ignored, because it was widely believed that the (non-Abelian gauge) theories in question were a dead-end, and in particular that they could not be renormalised. In 1971–72, Martinus Veltman and Gerard 't Hooft proved renormalisation of Yang–Mills was possible in two papers covering massless, and then massive, fields.[71] Their contribution, and others' work on the renormalization group - including "substantial" theoretical work by Russian physicists Ludvig Faddeev, Andrei Slavnov, Efim Fradkin and Igor Tyutin[72] - was eventually "enormously profound and influential",[73] but even with all key elements of the eventual theory published there was still almost no wider interest. For example, Coleman found in a study that "essentially no-one paid any attention" to Weinberg's paper prior to 1971[74] and discussed by David Politzer in his 2004 Nobel speech.[73] – now the most cited in particle physics[75] – and even in 1970 according to Politzer, Glashow's teaching of the weak interaction contained no mention of Weinberg's, Salam's, or Glashow's own work.[73] In practice, Politzer states, almost everyone learned of the theory due to physicist Benjamin Lee, who combined the work of Veltman and 't Hooft with insights by others, and popularised the completed theory.[73] In this way, from 1971, interest and acceptance "exploded" [73] and the ideas were quickly absorbed in the mainstream.[71][73]

The resulting electroweak theory and Standard Model have accurately predicted (among other things) weak neutral currents, three bosons, the top and charm quarks, and with great precision, the mass and other properties of some of these.[Note 7] Many of those involved eventually won Nobel Prizes or other renowned awards. A 1974 paper and comprehensive review in *Reviews of Modern Physics* commented that "while no one doubted the [mathematical] correctness of these arguments, no one quite believed that nature was diabolically clever enough to take advantage of them",[76]:9 adding that the theory had so far produced accurate answers that accorded with experiment, but it was unknown whether the theory was fundamentally correct.[76]:9,36(footnote),43–44,47 By 1986 and again in the 1990s it became possible to write that understanding and proving the Higgs sector of the Standard Model was "the central problem today in particle physics".[8][9]

7.3.1 Summary and impact of the PRL papers

The three papers written in 1964 were each recognised as milestone papers during *Physical Review Letters* 's 50th anniversary celebration.[67] Their six authors were also awarded the 2010 J. J. Sakurai Prize for Theoretical Particle Physics for this work.[77] (A controversy also arose the same year, because in the event of a Nobel Prize only up to three scientists could be recognised, with six being credited for the papers.[78]) Two of the three PRL papers (by Higgs and by GHK) contained equations for the hypothetical field that eventually would become known as the Higgs field and its hypothetical quantum, the Higgs boson.[59][60] Higgs' subsequent 1966 paper showed the decay mechanism of the boson; only a massive boson can decay and the decays can prove the mechanism.

In the paper by Higgs the boson is massive, and in a closing sentence Higgs writes that "an essential feature" of the theory "is the prediction of incomplete multiplets of scalar and vector bosons".[59] (Frank Close comments that 1960s gauge theorists were focused on the problem of massless *vector* bosons, and the implied existence of a massive *scalar* boson was not seen as important; only Higgs directly addressed it.[79]:154, 166, 175) In the paper by GHK the boson is massless and decoupled from the massive states.[60] In reviews dated 2009 and 2011, Guralnik states that in the GHK model the boson is massless only in a lowest-order approximation, but it is not subject to any constraint and acquires mass at higher orders, and adds that the GHK paper was the only one to show that there are no massless Goldstone bosons in the model and to give a complete analysis of the general Higgs mechanism.[50][80] All three reached similar conclusions, despite their very different approaches: Higgs' paper essentially used classical techniques, Englert and Brout's involved calculating vacuum polarization in perturbation theory around an assumed symmetry-breaking vacuum state, and GHK used operator formalism and conservation laws to explore in depth the ways in which Goldstone's theorem may be worked around.[51]

7.4 Theoretical properties

Main article: Higgs mechanism

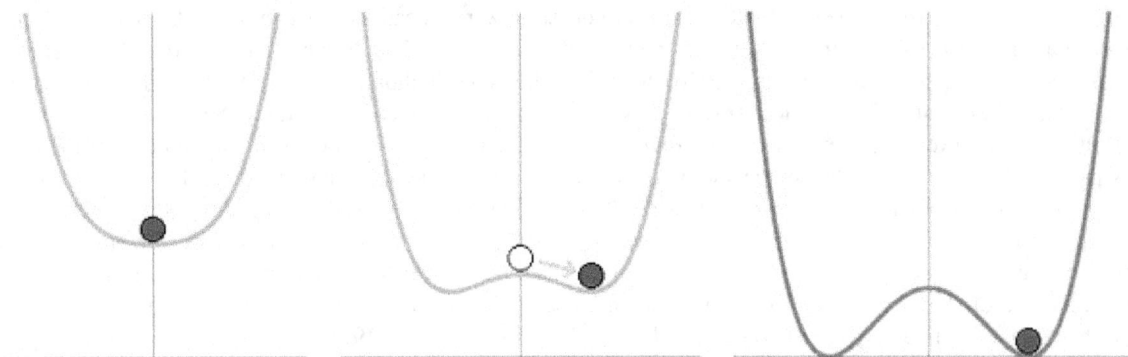

"Symmetry breaking illustrated": – At high energy levels (left) *the ball settles in the center, and the result is symmetrical. At lower energy levels* (right), *the overall "rules" remain symmetrical, but the "Mexican hat" potential comes into effect: "local" symmetry inevitably becomes broken since eventually the ball must at random roll one way or another.*

7.4.1   Theoretical need for the Higgs

Gauge invariance is an important property of modern particle theories such as the Standard Model, partly due to its success in other areas of fundamental physics such as electromagnetism and the strong interaction (quantum chromodynamics). However, there were great difficulties in developing gauge theories for the weak nuclear force or a possible unified electroweak interaction. Fermions with a mass term would violate gauge symmetry and therefore cannot be gauge invariant. (This can be seen by examining the Dirac Lagrangian for a fermion in terms of left and right handed components; we find none of the spin-half particles could ever flip helicity as required for mass, so they must be massless.[Note 13]) W and Z bosons are observed to have mass, but a boson mass term contains terms, which clearly depend on the choice of gauge and therefore these masses too cannot be gauge invariant. Therefore, it seems that *none* of the standard model fermions *or* bosons could "begin" with mass as an inbuilt property except by abandoning gauge invariance. If gauge invariance were to be retained, then these particles had to be acquiring their mass by some other mechanism or interaction. Additionally, whatever was giving these particles their mass, had to not "break" gauge invariance as the basis for other parts of the theories where it worked well, *and* had to not require or predict unexpected massless particles and long-range forces (seemingly an inevitable consequence of Goldstone's theorem) which did not actually seem to exist in nature.

A solution to all of these overlapping problems came from the discovery of a previously unnoticed borderline case hidden in the mathematics of Goldstone's theorem,[Note 11] that under certain conditions it *might* theoretically be possible for a symmetry to be broken *without* disrupting gauge invariance and *without* any new massless particles or forces, and having "sensible" (renormalisable) results mathematically: this became known as the Higgs mechanism.

The Standard Model hypothesizes a field which is responsible for this effect, called the Higgs field (symbol: ϕ), which has the unusual property of a non-zero amplitude in its ground state; i.e., a non-zero vacuum expectation value. It can have this effect because of its unusual "Mexican hat" shaped potential whose lowest "point" is not at its "centre". Below a certain extremely high energy level the existence of this non-zero vacuum expectation spontaneously breaks electroweak gauge symmetry which in turn gives rise to the Higgs mechanism and triggers the acquisition of mass by those particles interacting with the field. This effect occurs because scalar field components of the Higgs field are "absorbed" by the massive bosons as degrees of freedom, and couple to the fermions via Yukawa coupling, thereby producing the expected mass terms. In effect when symmetry breaks under these conditions, the Goldstone bosons that arise *interact* with the Higgs field (and with other particles capable of interacting with the Higgs field) instead of becoming new massless particles, the intractable problems of both underlying theories "neutralise" each other, and the residual outcome is that elementary particles acquire a consistent mass based on how strongly they interact with the Higgs field. It is the simplest known process capable of giving mass to the gauge bosons while remaining compatible with gauge theories.[81] Its quantum would be a scalar boson, known as the Higgs boson.[82]

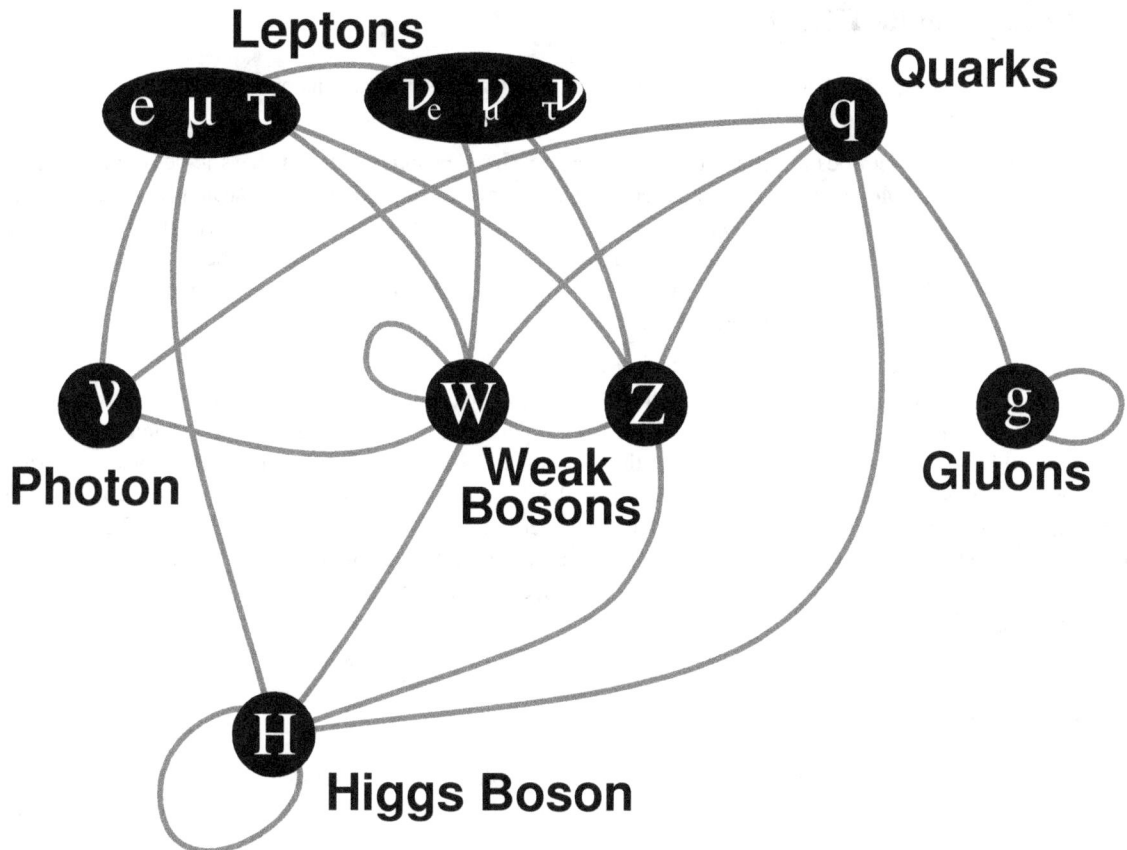

Summary of interactions between certain particles described by the Standard Model.

7.4.2 Properties of the Higgs field

In the Standard Model, the Higgs field is a scalar tachyonic field – 'scalar' meaning it does not transform under Lorentz transformations, and 'tachyonic' meaning the field (but not the particle) has imaginary mass and in certain configurations must undergo symmetry breaking. It consists of four components, two neutral ones and two charged component fields. Both of the charged components and one of the neutral fields are Goldstone bosons, which act as the longitudinal third-polarization components of the massive W^+, W^-, and Z bosons. The quantum of the remaining neutral component corresponds to (and is theoretically realised as) the massive Higgs boson,[83] this component can interact with fermions via Yukawa coupling to give them mass, as well.

Mathematically, the Higgs field has imaginary mass and is therefore a tachyonic field.[84] While tachyons (particles that move faster than light) are a purely hypothetical concept, fields with imaginary mass have come to play an important role in modern physics.[85][86] Under no circumstances do any excitations ever propagate faster than light in such theories — the presence or absence of a tachyonic mass has no effect whatsoever on the maximum velocity of signals (there is no violation of causality).[87] Instead of faster-than-light particles, the imaginary mass creates an instability:- any configuration in which one or more field excitations are tachyonic must spontaneously decay, and the resulting configuration contains no physical tachyons. This process is known as tachyon condensation, and is now believed to be the explanation for how the Higgs mechanism itself arises in nature, and therefore the reason behind electroweak symmetry breaking.

Although the notion of imaginary mass might seem troubling, it is only the field, and not the mass itself, that is quantized. Therefore, the field operators at spacelike separated points still commute (or anticommute), and information and particles still do not propagate faster than light.[88] Tachyon condensation drives a physical system that has reached a local limit and might naively be expected to produce physical tachyons, to an alternate stable state where no physical tachyons exist. Once a tachyonic field such as the Higgs field reaches the minimum of the potential, its quanta are not tachyons any more but rather are ordinary particles such as the Higgs boson.[89]

7.4.3 Properties of the Higgs boson

Since the Higgs field is scalar, the Higgs boson has no spin. The Higgs boson is also its own antiparticle and is CP-even, and has zero electric and colour charge.[90]

The Minimal Standard Model does not predict the mass of the Higgs boson.[91] If that mass is between 115 and 180 GeV/c^2, then the Standard Model can be valid at energy scales all the way up to the Planck scale (10^{19} GeV).[92] Many theorists expect new physics beyond the Standard Model to emerge at the TeV-scale, based on unsatisfactory properties of the Standard Model.[93] The highest possible mass scale allowed for the Higgs boson (or some other electroweak symmetry breaking mechanism) is 1.4 TeV; beyond this point, the Standard Model becomes inconsistent without such a mechanism, because unitarity is violated in certain scattering processes.[94]

It is also possible, although experimentally difficult, to estimate the mass of the Higgs boson indirectly. In the Standard Model, the Higgs boson has a number of indirect effects; most notably, Higgs loops result in tiny corrections to masses of W and Z bosons. Precision measurements of electroweak parameters, such as the Fermi constant and masses of W/Z bosons, can be used to calculate constraints on the mass of the Higgs. As of July 2011, the precision electroweak measurements tell us that the mass of the Higgs boson is likely to be less than about 161 GeV/c^2 at 95% confidence level (this upper limit would increase to 185 GeV/c^2 if the lower bound of 114.4 GeV/c^2 from the LEP-2 direct search is allowed for[95]). These indirect constraints rely on the assumption that the Standard Model is correct. It may still be possible to discover a Higgs boson above these masses if it is accompanied by other particles beyond those predicted by the Standard Model.[96]

7.4.4 Production

If Higgs particle theories are valid, then a Higgs particle can be produced much like other particles that are studied, in a particle collider. This involves accelerating a large number of particles to extremely high energies and extremely close to the speed of light, then allowing them to smash together. Protons and lead ions (the bare nuclei of lead atoms) are used at the LHC. In the extreme energies of these collisions, the desired esoteric particles will occasionally be produced and this can be detected and studied; any absence or difference from theoretical expectations can also be used to improve the theory. The relevant particle theory (in this case the Standard Model) will determine the necessary kinds of collisions and detectors. The Standard Model predicts that Higgs bosons could be formed in a number of ways,[97][98][99] although the probability of producing a Higgs boson in any collision is always expected to be very small—for example, only 1 Higgs boson per 10 billion collisions in the Large Hadron Collider.[Note 14] The most common expected processes for Higgs boson production are:

- *Gluon fusion.* If the collided particles are hadrons such as the proton or antiproton—as is the case in the LHC and Tevatron—then it is most likely that two of the gluons binding the hadron together collide. The easiest way to produce a Higgs particle is if the two gluons combine to form a loop of virtual quarks. Since the coupling of particles to the Higgs boson is proportional to their mass, this process is more likely for heavy particles. In practice it is enough to consider the contributions of virtual top and bottom quarks (the heaviest quarks). This process is the dominant contribution at the LHC and Tevatron being about ten times more likely than any of the other processes.[97][98]

- *Higgs Strahlung.* If an elementary fermion collides with an anti-fermion—e.g., a quark with an anti-quark or an electron with a positron—the two can merge to form a virtual W or Z boson which, if it carries sufficient energy, can then emit a Higgs boson. This process was the dominant production mode at the LEP, where an electron and a positron collided to form a virtual Z boson, and it was the second largest contribution for Higgs production at the Tevatron. At the LHC this process is only the third largest, because the LHC collides protons with protons, making a quark-antiquark collision less likely than at the Tevatron. Higgs Strahlung is also known as *associated production.*[97][98][99]

- *Weak boson fusion.* Another possibility when two (anti-)fermions collide is that the two exchange a virtual W or Z boson, which emits a Higgs boson. The colliding fermions do not need to be the same type. So, for example, an up quark may exchange a Z boson with an anti-down quark. This process is the second most important for the production of Higgs particle at the LHC and LEP.[97][99]

- *Top fusion*. The final process that is commonly considered is by far the least likely (by two orders of magnitude). This process involves two colliding gluons, which each decay into a heavy quark–antiquark pair. A quark and antiquark from each pair can then combine to form a Higgs particle.[97][98]

7.4.5 Decay

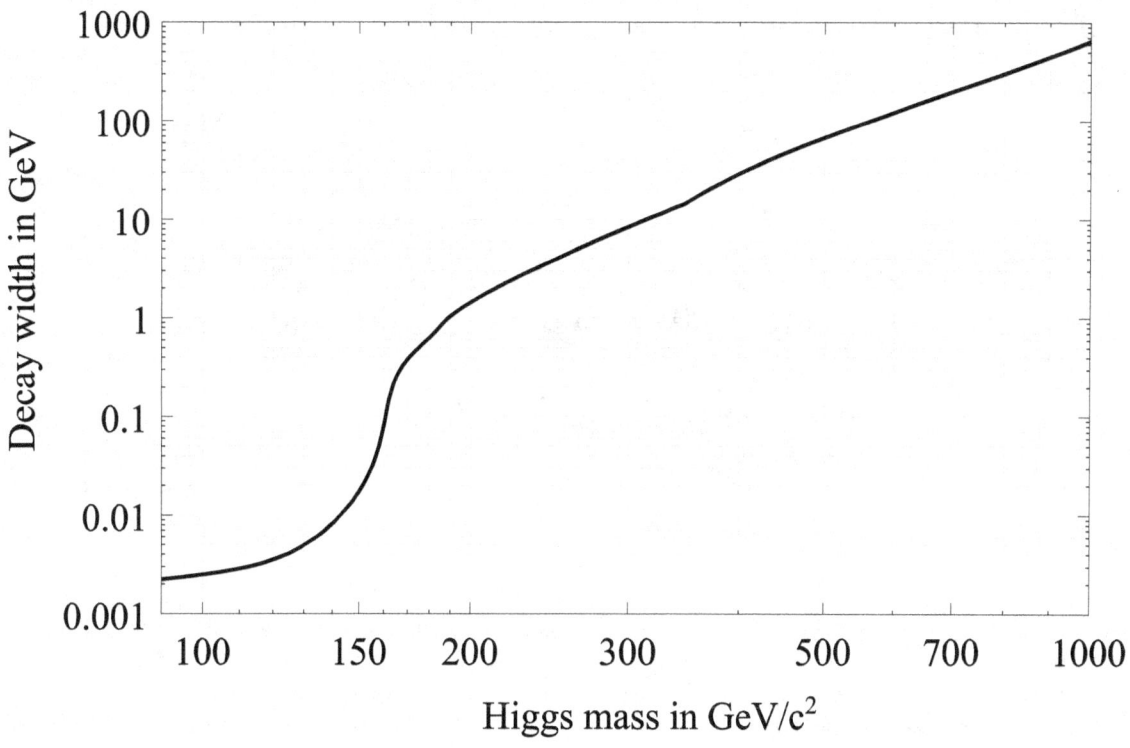

The Standard Model prediction for the decay width of the Higgs particle depends on the value of its mass.

Quantum mechanics predicts that if it is possible for a particle to decay into a set of lighter particles, then it will eventually do so.[101] This is also true for the Higgs boson. The likelihood with which this happens depends on a variety of factors including: the difference in mass, the strength of the interactions, etc. Most of these factors are fixed by the Standard Model, except for the mass of the Higgs boson itself. For a Higgs boson with a mass of 126 GeV/c^2 the SM predicts a mean life time of about 1.6×10^{-22} s.[Note 2]

Since it interacts with all the massive elementary particles of the SM, the Higgs boson has many different processes through which it can decay. Each of these possible processes has its own probability, expressed as the *branching ratio*; the fraction of the total number decays that follows that process. The SM predicts these branching ratios as a function of the Higgs mass (see plot).

One way that the Higgs can decay is by splitting into a fermion–antifermion pair. As general rule, the Higgs is more likely to decay into heavy fermions than light fermions, because the mass of a fermion is proportional to the strength of its interaction with the Higgs.[102] By this logic the most common decay should be into a top–antitop quark pair. However, such a decay is only possible if the Higgs is heavier than ~346 GeV/c^2, twice the mass of the top quark. For a Higgs mass of 126 GeV/c^2 the SM predicts that the most common decay is into a bottom–antibottom quark pair, which happens 56.1% of the time.[5] The second most common fermion decay at that mass is a tau–antitau pair, which happens only about 6% of the time.[5]

Another possibility is for the Higgs to split into a pair of massive gauge bosons. The most likely possibility is for the Higgs to decay into a pair of W bosons (the light blue line in the plot), which happens about 23.1% of the time for a

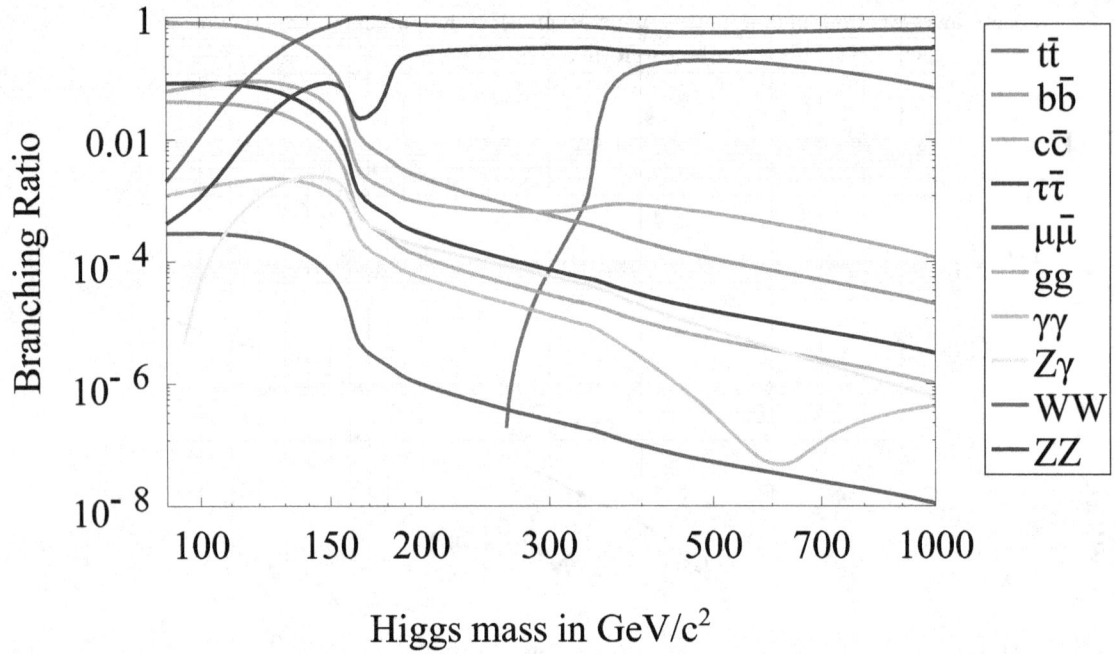

The Standard Model prediction for the branching ratios of the different decay modes of the Higgs particle depends on the value of its mass.

Higgs boson with a mass of 126 GeV/c^2.[5] The W bosons can subsequently decay either into a quark and an antiquark or into a charged lepton and a neutrino. However, the decays of W bosons into quarks are difficult to distinguish from the background, and the decays into leptons cannot be fully reconstructed (because neutrinos are impossible to detect in particle collision experiments). A cleaner signal is given by decay into a pair of Z-bosons (which happens about 2.9% of the time for a Higgs with a mass of 126 GeV/c^2),[5] if each of the bosons subsequently decays into a pair of easy-to-detect charged leptons (electrons or muons).

Decay into massless gauge bosons (i.e., gluons or photons) is also possible, but requires intermediate loop of virtual heavy quarks (top or bottom) or massive gauge bosons.[102] The most common such process is the decay into a pair of gluons through a loop of virtual heavy quarks. This process, which is the reverse of the gluon fusion process mentioned above, happens approximately 8.5% of the time for a Higgs boson with a mass of 126 GeV/c^2.[5] Much rarer is the decay into a pair of photons mediated by a loop of W bosons or heavy quarks, which happens only twice for every thousand decays.[5] However, this process is very relevant for experimental searches for the Higgs boson, because the energy and momentum of the photons can be measured very precisely, giving an accurate reconstruction of the mass of the decaying particle.[102]

7.4.6   Alternative models

Main article: Alternatives to the Standard Model Higgs

The Minimal Standard Model as described above is the simplest known model for the Higgs mechanism with just one Higgs field. However, an extended Higgs sector with additional Higgs particle doublets or triplets is also possible, and many extensions of the Standard Model have this feature. The non-minimal Higgs sector favoured by theory are the two-Higgs-doublet models (2HDM), which predict the existence of a quintet of scalar particles: two CP-even neutral Higgs bosons h^0 and H^0, a CP-odd neutral Higgs boson A^0, and two charged Higgs particles $H^\pm$. Supersymmetry ("SUSY") also predicts relations between the Higgs-boson masses and the masses of the gauge bosons, and could accommodate a 125 GeV/c^2 neutral Higgs boson.

The key method to distinguish between these different models involves study of the particles' interactions ("coupling") and exact decay processes ("branching ratios"), which can be measured and tested experimentally in particle collisions. In the Type-I 2HDM model one Higgs doublet couples to up and down quarks, while the second doublet does not couple to quarks. This model has two interesting limits, in which the lightest Higgs couples to just fermions ("gauge-phobic") or just gauge bosons ("fermiophobic"), but not both. In the Type-II 2HDM model, one Higgs doublet only couples to up-type quarks, the other only couples to down-type quarks.[103] The heavily researched Minimal Supersymmetric Standard Model (MSSM) includes a Type-II 2HDM Higgs sector, so it could be disproven by evidence of a Type-I 2HDM Higgs.

In other models the Higgs scalar is a composite particle. For example, in technicolor the role of the Higgs field is played by strongly bound pairs of fermions called techniquarks. Other models, feature pairs of top quarks (see top quark condensate). In yet other models, there is no Higgs field at all and the electroweak symmetry is broken using extra dimensions.[104][105]

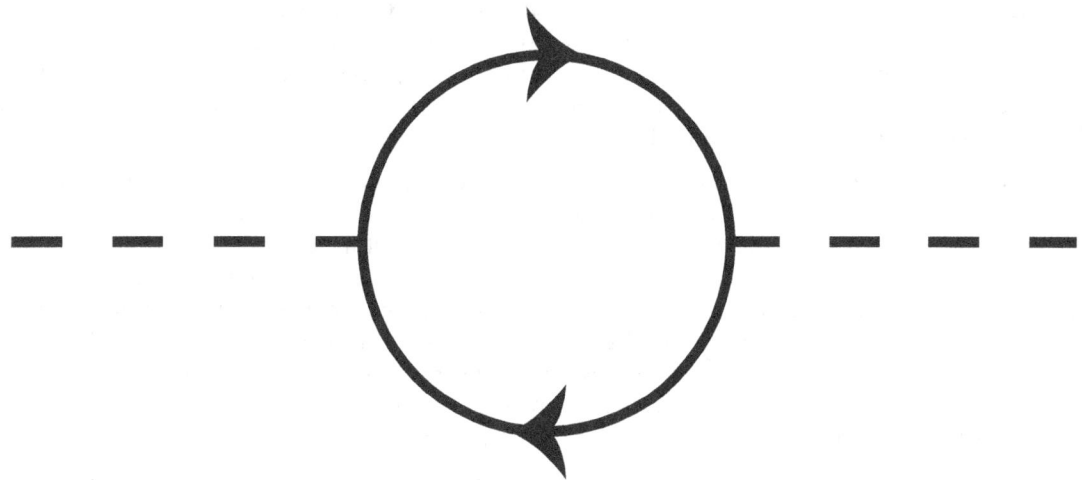

A one-loop Feynman diagram of the first-order correction to the Higgs mass. In the Standard Model the effects of these corrections are potentially enormous, giving rise to the so-called hierarchy problem.

7.4.7 Further theoretical issues and hierarchy problem

Main articles: Hierarchy problem and Hierarchy problem § The Higgs mass

The Standard Model leaves the mass of the Higgs boson as a parameter to be measured, rather than a value to be calculated. This is seen as theoretically unsatisfactory, particularly as quantum corrections (related to interactions with virtual particles) should apparently cause the Higgs particle to have a mass immensely higher than that observed, but at the same time the Standard Model requires a mass of the order of 100 to 1000 GeV to ensure unitarity (in this case, to unitarise longitudinal vector boson scattering).[106] Reconciling these points appears to require explaining why there is an almost-perfect cancellation resulting in the visible mass of ~ 125 GeV, and it is not clear how to do this. Because the weak force is about 10^{32} times stronger than gravity, and (linked to this) the Higgs boson's mass is so much less than the Planck mass or the grand unification energy, it appears that either there is some underlying connection or reason for these observations which is unknown and not described by the Standard Model, or some unexplained and extremely precise fine-tuning of parameters – however at present neither of these explanations is proven. This is known as a hierarchy problem.[107] More broadly, the hierarchy problem amounts to the worry that a future theory of fundamental particles and interactions should not have excessive fine-tunings or unduly delicate cancellations, and should allow masses of particles such as the Higgs boson to be calculable. The problem is in some ways unique to spin-0 particles (such as the Higgs boson), which can give rise to issues related to quantum corrections that do not affect particles with spin.[106] A number of solutions have been proposed, including supersymmetry, conformal solutions and solutions via extra dimensions such as braneworld models.

There are also issues of quantum triviality, which suggests that it may not be possible to create a consistent quantum field theory involving elementary scalar particles.

7.5   Experimental search

Main article: Search for the Higgs boson

To produce Higgs bosons, two beams of particles are accelerated to very high energies and allowed to collide within a particle detector. Occasionally, although rarely, a Higgs boson will be created fleetingly as part of the collision byproducts. Because the Higgs boson decays very quickly, particle detectors cannot detect it directly. Instead the detectors register all the decay products (the *decay signature*) and from the data the decay process is reconstructed. If the observed decay products match a possible decay process (known as a *decay channel*) of a Higgs boson, this indicates that a Higgs boson may have been created. In practice, many processes may produce similar decay signatures. Fortunately, the Standard Model precisely predicts the likelihood of each of these, and each known process, occurring. So, if the detector detects more decay signatures consistently matching a Higgs boson than would otherwise be expected if Higgs bosons did not exist, then this would be strong evidence that the Higgs boson exists.

Because Higgs boson production in a particle collision is likely to be very rare (1 in 10 billion at the LHC),[Note 14] and many other possible collision events can have similar decay signatures, the data of hundreds of trillions of collisions needs to be analysed and must "show the same picture" before a conclusion about the existence of the Higgs boson can be reached. To conclude that a new particle has been found, particle physicists require that the statistical analysis of two independent particle detectors each indicate that there is lesser than a one-in-a-million chance that the observed decay signatures are due to just background random Standard Model events—i.e., that the observed number of events is more than 5 standard deviations (sigma) different from that expected if there was no new particle. More collision data allows better confirmation of the physical properties of any new particle observed, and allows physicists to decide whether it is indeed a Higgs boson as described by the Standard Model or some other hypothetical new particle.

To find the Higgs boson, a powerful particle accelerator was needed, because Higgs bosons might not be seen in lower-energy experiments. The collider needed to have a high luminosity in order to ensure enough collisions were seen for conclusions to be drawn. Finally, advanced computing facilities were needed to process the vast amount of data (25 petabytes per year as at 2012) produced by the collisions.[109] For the announcement of 4 July 2012, a new collider known as the Large Hadron Collider was constructed at CERN with a planned eventual collision energy of 14 TeV—over seven times any previous collider—and over 300 trillion (3×10^{14}) LHC proton–proton collisions were analysed by the LHC Computing Grid, the world's largest computing grid (as of 2012), comprising over 170 computing facilities in a worldwide network across 36 countries.[109][110][111]

7.5.1   Search prior to 4 July 2012

The first extensive search for the Higgs boson was conducted at the Large Electron–Positron Collider (LEP) at CERN in the 1990s. At the end of its service in 2000, LEP had found no conclusive evidence for the Higgs.[Note 15] This implied that if the Higgs boson were to exist it would have to be heavier than 114.4 GeV/c^2.[112]

The search continued at Fermilab in the United States, where the Tevatron—the collider that discovered the top quark in 1995—had been upgraded for this purpose. There was no guarantee that the Tevatron would be able to find the Higgs, but it was the only supercollider that was operational since the Large Hadron Collider (LHC) was still under construction and the planned Superconducting Super Collider had been cancelled in 1993 and never completed. The Tevatron was only able to exclude further ranges for the Higgs mass, and was shut down on 30 September 2011 because it no longer could keep up with the LHC. The final analysis of the data excluded the possibility of a Higgs boson with a mass between 147 GeV/c^2 and 180 GeV/c^2. In addition, there was a small (but not significant) excess of events possibly indicating a Higgs boson with a mass between 115 GeV/c^2 and 140 GeV/c^2.[113]

The Large Hadron Collider at CERN in Switzerland, was designed specifically to be able to either confirm or exclude the existence of the Higgs boson. Built in a 27 km tunnel under the ground near Geneva originally inhabited by LEP, it was designed to collide two beams of protons, initially at energies of 3.5 TeV per beam (7 TeV total), or almost 3.6 times that of the Tevatron, and upgradeable to 2 × 7 TeV (14 TeV total) in future. Theory suggested if the Higgs boson existed, collisions at these energy levels should be able to reveal it. As one of the most complicated scientific instruments ever built, its operational readiness was delayed for 14 months by a magnet quench event nine days after its inaugural tests, caused by a faulty electrical connection that damaged over 50 superconducting magnets and contaminated the vacuum system.[114][115][116]

Data collection at the LHC finally commenced in March 2010.[117] By December 2011 the two main particle detectors at the LHC, ATLAS and CMS, had narrowed down the mass range where the Higgs could exist to around 116-130 GeV (ATLAS) and 115-127 GeV (CMS).[118][119] There had also already been a number of promising event excesses that had "evaporated" and proven to be nothing but random fluctuations. However, from around May 2011,[120] both experiments had seen among their results, the slow emergence of a small yet consistent excess of gamma and 4-lepton decay signatures and several other particle decays, all hinting at a new particle at a mass around 125 GeV.[120] By around November 2011, the anomalous data at 125 GeV was becoming "too large to ignore" (although still far from conclusive), and the team leaders at both ATLAS and CMS each privately suspected they might have found the Higgs.[120] On November 28, 2011, at an internal meeting of the two team leaders and the director general of CERN, the latest analyses were discussed outside their teams for the first time, suggesting both ATLAS and CMS might be converging on a possible shared result at 125 GeV, and initial preparations commenced in case of a successful finding.[120] While this information was not known publicly at the time, the narrowing of the possible Higgs range to around 115–130 GeV and the repeated observation of small but consistent event excesses across multiple channels at both ATLAS and CMS in the 124-126 GeV region (described as "tantalising hints" of around 2-3 sigma) were public knowledge with "a lot of interest".[121] It was therefore widely anticipated around the end of 2011, that the LHC would provide sufficient data to either exclude or confirm the finding of a Higgs boson by the end of 2012, when their 2012 collision data (with slightly higher 8 TeV collision energy) had been examined.[121][122]

7.5.2 Discovery of candidate boson at CERN

On 22 June 2012 CERN announced an upcoming seminar covering tentative findings for 2012,[126][127] and shortly afterwards (from around 1 July 2012 according to an analysis of the spreading rumour in social media[128]) rumours began to spread in the media that this would include a major announcement, but it was unclear whether this would be a stronger signal or a formal discovery.[129][130] Speculation escalated to a "fevered" pitch when reports emerged that Peter Higgs, who proposed the particle, was to be attending the seminar,[131][132] and that "five leading physicists" had been invited – generally believed to signify the five living 1964 authors – with Higgs, Englert, Guralnik, Hagen attending and Kibble confirming his invitation (Brout having died in 2011).[133][134]

On 4 July 2012 both of the CERN experiments announced they had independently made the same discovery:[135] CMS of a previously unknown boson with mass 125.3 ± 0.6 GeV/c^2[136][137] and ATLAS of a boson with mass 126.0 ± 0.6 GeV/c^2.[138][139] Using the combined analysis of two interaction types (known as 'channels'), both experiments independently reached a local significance of 5 sigma — implying that the probability of getting at least as strong a result by chance alone is less than 1 in 3 million. When additional channels were taken into account, the CMS significance was reduced to 4.9 sigma.[137]

The two teams had been working 'blinded' from each other from around late 2011 or early 2012,[120] meaning they did not discuss their results with each other, providing additional certainty that any common finding was genuine validation of a particle.[109] This level of evidence, confirmed independently by two separate teams and experiments, meets the formal level of proof required to announce a confirmed discovery.

On 31 July 2012, the ATLAS collaboration presented additional data analysis on the "observation of a new particle", including data from a third channel, which improved the significance to 5.9 sigma (1 in 588 million chance of obtaining at least as strong evidence by random background effects alone) and mass 126.0 ± 0.4 (stat) ± 0.4 (sys) GeV/c^2, [139] and CMS improved the significance to 5-sigma and mass 125.3 ± 0.4 (stat) ± 0.5 (sys) GeV/c^2.[136]

7.5.3 The new particle tested as a possible Higgs boson

Following the 2012 discovery, it was still unconfirmed whether or not the 125 GeV/c^2 particle was a Higgs boson. On one hand, observations remained consistent with the observed particle being the Standard Model Higgs boson, and the particle decayed into at least some of the predicted channels. Moreover, the production rates and branching ratios for the observed channels broadly matched the predictions by the Standard Model within the experimental uncertainties. However, the experimental uncertainties currently still left room for alternative explanations, meaning an announcement of the discovery of a Higgs boson would have been premature.[102] To allow more opportunity for data collection, the LHC's proposed 2012 shutdown and 2013–14 upgrade were postponed by 7 weeks into 2013.[140]

In November 2012, in a conference in Kyoto researchers said evidence gathered since July was falling into line with the basic Standard Model more than its alternatives, with a range of results for several interactions matching that theory's predictions.[141] Physicist Matt Strassler highlighted "considerable" evidence that the new particle is not a pseudoscalar negative parity particle (consistent with this required finding for a Higgs boson), "evaporation" or lack of increased significance for previous hints of non-Standard Model findings, expected Standard Model interactions with W and Z bosons, absence of "significant new implications" for or against supersymmetry, and in general no significant deviations to date from the results expected of a Standard Model Higgs boson.[142] However some kinds of extensions to the Standard Model would also show very similar results;[143] so commentators noted that based on other particles that are still being understood long after their discovery, it may take years to be sure, and decades to fully understand the particle that has been found.[141][142]

These findings meant that as of January 2013, scientists were very sure they had found an unknown particle of mass ~ 125 GeV/c^2, and had not been misled by experimental error or a chance result. They were also sure, from initial observations, that the new particle was some kind of boson. The behaviours and properties of the particle, so far as examined since July 2012, also seemed quite close to the behaviours expected of a Higgs boson. Even so, it could still have been a Higgs boson or some other unknown boson, since future tests could show behaviours that do not match a Higgs boson, so as of December 2012 CERN still only stated that the new particle was "consistent with" the Higgs boson,[11][13] and scientists did not yet positively say it was the Higgs boson.[144] Despite this, in late 2012, widespread media reports announced (incorrectly) that a Higgs boson had been confirmed during the year.[Note 16]

In January 2013, CERN director-general Rolf-Dieter Heuer stated that based on data analysis to date, an answer could be possible 'towards' mid-2013,[150] and the deputy chair of physics at Brookhaven National Laboratory stated in February 2013 that a "definitive" answer might require "another few years" after the collider's 2015 restart.[151] In early March 2013, CERN Research Director Sergio Bertolucci stated that confirming spin-0 was the major remaining requirement to determine whether the particle is at least some kind of Higgs boson.[152]

7.5.4   Preliminary confirmation of existence and current status

On 14 March 2013 CERN confirmed that:

> "CMS and ATLAS have compared a number of options for the spin-parity of this particle, and these all prefer no spin and even parity [two fundamental criteria of a Higgs boson consistent with the Standard Model]. This, coupled with the measured interactions of the new particle with other particles, strongly indicates that it is a Higgs boson." [1]

This also makes the particle the first elementary scalar particle to be discovered in nature.[14]

Examples of tests used to validate whether the 125 GeV particle is a Higgs boson:[142][153]

7.6   Public discussion

7.6.1   Naming

Names used by physicists

The name most strongly associated with the particle and field is the Higgs boson[79]:168 and Higgs field. For some time the particle was known by a combination of its PRL author names (including at times Anderson), for example the Brout–Englert–Higgs particle, the Anderson-Higgs particle, or the Englert–Brout–Higgs–Guralnik–Hagen–Kibble mechanism,[Note 17] and these are still used at times.[51][160] Fueled in part by the issue of recognition and a potential shared Nobel Prize,[160][161] the most appropriate name is still occasionally a topic of debate as at 2012.[160] (Higgs himself prefers to call the particle either by an acronym of all those involved, or "the scalar boson", or "the so-called Higgs particle".[161])

A considerable amount has been written on how Higgs' name came to be exclusively used. Two main explanations are offered.

Nickname

The Higgs boson is often referred to as the "God particle" in popular media outside the scientific community.[170][171][172][The nickname comes from the title of the 1993 book on the Higgs boson and particle physics - The God Particle: If the Universe Is the Answer,What Is the Question? byNobel Physics prizewinnerandFermilabdirectorLeon Lederman. [21] Lederman wrote it in the context of failing US government support for the Superconducting Super Collider,[175] a part-constructed titanic[176][177] competitor to the Large Hadron Collider with planned collision energies of 2×20 TeV that was championed by Lederman since its 1983 inception[175][178][179] and shut down in 1993. The book sought in part to promote awareness of the significance and need for such a project in the face of its possible loss of funding.[180] Lederman, a leading researcher in the field, wanted to title his book "The Goddamn Particle: If the Universe is the Answer, What is the Question?" But his editor decided that the title was too controversial and convinced Lederman to change the title to "The God Particle: If the Universe is the Answer, What is the Question?"[181]

And since the Higgs Boson deals with how matter was formed at the time of the big bang, and since newspapers loved the term, the term "God particle" was used.

While media use of this term may have contributed to wider awareness and interest,[182] many scientists feel the name is inappropriate[16][17][183] since it is sensational hyperbole and misleads readers;[184] the particle also has nothing to do with God, leaves open numerous questions in fundamental physics, and does not explain the ultimate origin of the universe. Higgs, an atheist, was reported to be displeased and stated in a 2008 interview that he found it "embarrassing" because it was "the kind of misuse... which I think might offend some people".[184][185][186] Science writer Ian Sample stated in his 2010 book on the search that the nickname is "universally hate[d]" by physicists and perhaps the "worst derided" in the history of physics, but that (according to Lederman) the publisher rejected all titles mentioning "Higgs" as unimaginative and too unknown.[187]

Lederman begins with a review of the long human search for knowledge, and explains that his tongue-in-cheek title draws an analogy between the impact of the Higgs field on the fundamental symmetries at the Big Bang, and the apparent chaos of structures, particles, forces and interactions that resulted and shaped our present universe, with the biblical story of Babel in which the primordial single language of early Genesis was fragmented into many disparate languages and cultures.[188]

> Today ... we have the standard model, which reduces all of reality to a dozen or so particles and four forces. ... It's a hard-won simplicity [...and...] remarkably accurate. But it is also incomplete and, in fact, internally inconsistent... This boson is so central to the state of physics today, so crucial to our final understanding of the structure of matter, yet so elusive, that I have given it a nickname: the God Particle. Why God Particle? Two reasons. One, the publisher wouldn't let us call it the Goddamn Particle, though that might be a more appropriate title, given its villainous nature and the expense it is causing. And two, there is a connection, of sorts, to another book, a *much* older one...
> — Leon M. Lederman and Dick Teresi, *The God Particle: If the Universe is the Answer, What is the Question*[21] p. 22

Lederman asks whether the Higgs boson was added just to perplex and confound those seeking knowledge of the universe, and whether physicists will be confounded by it as recounted in that story, or ultimately surmount the challenge and understand "how beautiful is the universe [God has] made".[189]

Other proposals

A renaming competition by British newspaper *The Guardian* in 2009 resulted in their science correspondent choosing the name "the champagne bottle boson" as the best submission: "The bottom of a champagne bottle is in the shape of the Higgs potential and is often used as an illustration in physics lectures. So it's not an embarrassingly grandiose name, it is memorable, and [it] has some physics connection too."[190] The name *Higgson* was suggested as well, in an opinion piece in the Institute of Physics' online publication *physicsworld.com*.[191]

7.6.2   Media explanations and analogies

There has been considerable public discussion of analogies and explanations for the Higgs particle and how the field creates mass,[192][193] including coverage of explanatory attempts in their own right and a competition in 1993 for the best popular explanation by then-UK Minister for Science Sir William Waldegrave[194] and articles in newspapers worldwide.

An educational collaboration involving an LHC physicist and a High School Teachers at CERN educator suggests that dispersion of light – responsible for the rainbow and dispersive prism – is a useful analogy for the Higgs field's symmetry breaking and mass-causing effect.[195]

Matt Strassler uses electric fields as an analogy:[196]

> Some particles interact with the Higgs field while others don't. Those particles that feel the Higgs field act as if they have mass. Something similar happens in an electric field – charged objects are pulled around and neutral objects can sail through unaffected. So you can think of the Higgs search as an attempt to make waves in the Higgs field *[create Higgs bosons]* to prove it's really there.

A similar explanation was offered by *The Guardian*:[197]

> The Higgs boson is essentially a ripple in a field said to have emerged at the birth of the universe and to span the cosmos to this day ... The particle is crucial however: it is the smoking gun, the evidence required to show the theory is right.

The Higgs field's effect on particles was famously described by physicist David Miller as akin to a room full of political party workers spread evenly throughout a room: the crowd gravitates to and slows down famous people but does not slow down others.[Note 18] He also drew attention to well-known effects in solid state physics where an electron's effective mass can be much greater than usual in the presence of a crystal lattice.[198]

Analogies based on drag effects, including analogies of "syrup" or "molasses" are also well known, but can be somewhat misleading since they may be understood (incorrectly) as saying that the Higgs field simply resists some particles' motion but not others' – a simple resistive effect could also conflict with Newton's third law.[200]

7.6.3   Recognition and awards

There has been considerable discussion of how to allocate the credit if the Higgs boson is proven, made more pointed as a Nobel prize had been expected, and the very wide basis of people entitled to consideration. These include a range of theoreticians who made the Higgs mechanism theory possible, the theoreticians of the 1964 PRL papers (including Higgs himself), the theoreticians who derived from these, a working electroweak theory and the Standard Model itself, and also the experimentalists at CERN and other institutions who made possible the proof of the Higgs field and boson in reality. The Nobel prize has a limit of 3 persons to share an award, and some possible winners are already prize holders for other work, or are deceased (the prize is only awarded to persons in their lifetime). Existing prizes for works relating to the Higgs field, boson, or mechanism include:

Photograph of light passing through a dispersive prism: the rainbow effect arises because photons are not all affected to the same degree by the dispersive material of the prism.

- Nobel Prize in Physics (1979) – Glashow, Salam, and Weinberg, *for contributions to the theory of the unified weak and electromagnetic interaction between elementary particles* [201]

- Nobel Prize in Physics (1999) – 't Hooft and Veltman, *for elucidating the quantum structure of electroweak inter-*

actions in physics [202]

- Nobel Prize in Physics (2008) – Nambu (shared), *for the discovery of the mechanism of spontaneous broken symmetry in subatomic physics* [53]

- J. J. Sakurai Prize for Theoretical Particle Physics (2010) – Hagen, Englert, Guralnik, Higgs, Brout, and Kibble, *for elucidation of the properties of spontaneous symmetry breaking in four-dimensional relativistic gauge theory and of the mechanism for the consistent generation of vector boson masses* [77] (for the 1964 papers described above)

- Wolf Prize (2004) – Englert, Brout, and Higgs

- Nobel Prize in Physics (2013) - Peter Higgs and François Englert, *for the theoretical discovery of a mechanism that contributes to our understanding of the origin of mass of subatomic particles, and which recently was confirmed through the discovery of the predicted fundamental particle, by the ATLAS and CMS experiments at CERN's Large Hadron Collider* [203]

Additionally Physical Review Letters' 50-year review (2008) recognized the 1964 PRL symmetry breaking papers and Weinberg's 1967 paper *A model of Leptons* (the most cited paper in particle physics, as of 2012) "milestone Letters".[75]

Following reported observation of the Higgs-like particle in July 2012, several Indian media outlets reported on the supposed neglect of credit to Indian physicist Satyendra Nath Bose after whose work in the 1920s the class of particles "bosons" is named[204][205] (although physicists have described Bose's connection to the discovery as tenuous).[206]

7.7   Technical aspects and mathematical formulation

See also: Standard Model (mathematical formulation)

In the Standard Model, the Higgs field is a four-component scalar field that forms a complex doublet of the weak isospin SU(2) symmetry:

while the field has charge +1/2 under the weak hypercharge U(1) symmetry (in the convention where the electric charge, Q, the weak isospin, I_3, and the weak hypercharge, Y, are related by $Q = I_3 + Y$).[207]

The Higgs part of the Lagrangian is[207]

where W_μ^a and B_μ are the gauge bosons of the SU(2) and U(1) symmetries, g and g' their respective coupling constants, $\tau^a = \sigma^a/2$ (where σ^a are the Pauli matrices) a complete set generators of the SU(2) symmetry, and $\lambda > 0$ and $\mu^2 > 0$, so that the ground state breaks the SU(2) symmetry (see figure). The ground state of the Higgs field (the bottom of the potential) is degenerate with different ground states related to each other by a SU(2) gauge transformation. It is always possible to pick a gauge such that in the ground state $\phi^1 = \phi^2 = \phi^3 = 0$. The expectation value of ϕ^0 in the ground state (the vacuum expectation value or vev) is then $\langle\phi^0\rangle = v$, where $v = \frac{|\mu|}{\sqrt{\lambda}}$. The measured value of this parameter is ~246 GeV/c^2.[102] It has units of mass, and is the only free parameter of the Standard Model that is not a dimensionless number. Quadratic terms in W_μ and B_μ arise, which give masses to the W and Z bosons:[207]

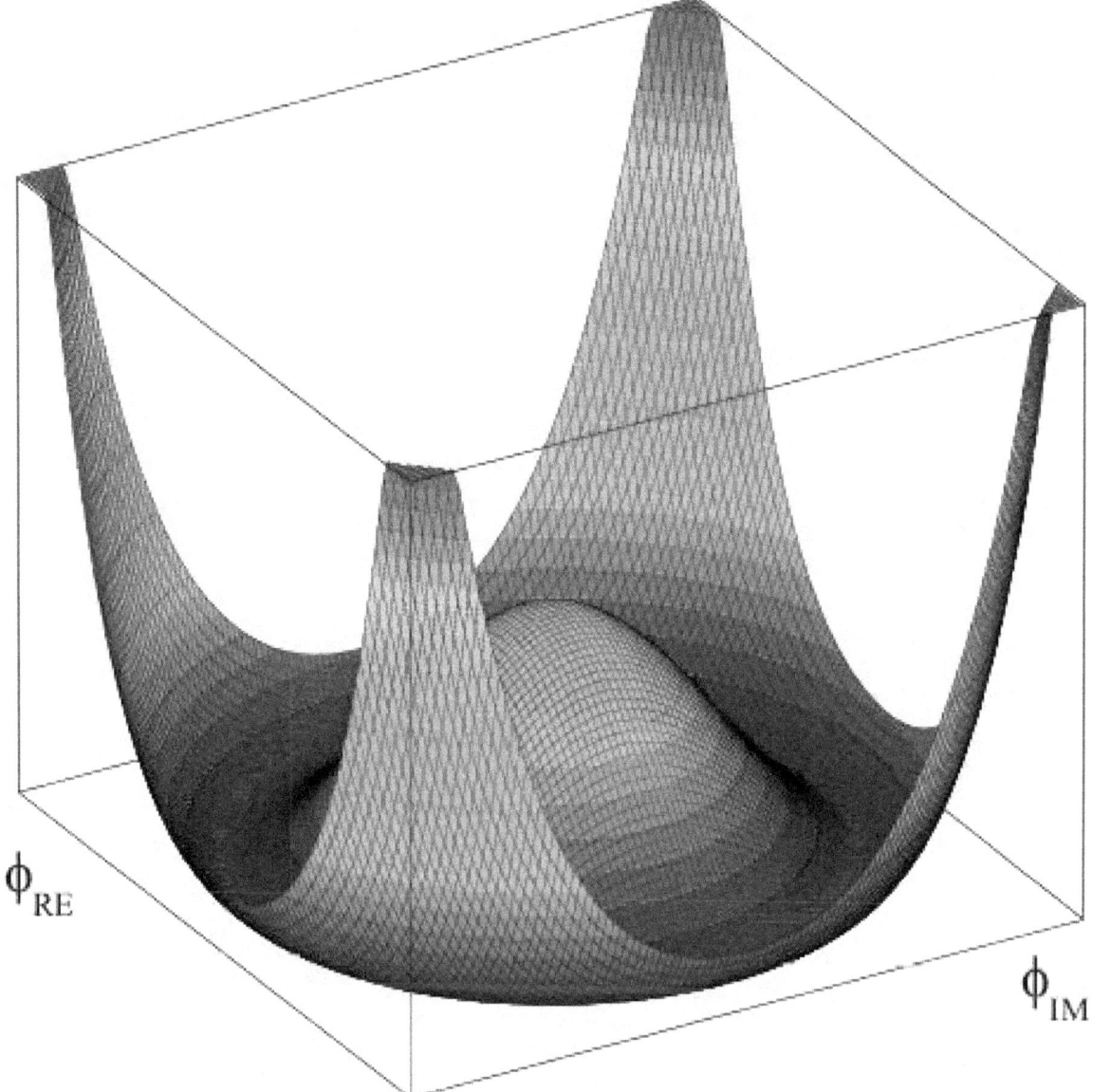

The potential for the Higgs field, plotted as function of ϕ^0 and ϕ^3. It has a Mexican-hat *or* champagne-bottle profile *at the ground.*

with their ratio determining the Weinberg angle, $\cos\theta_W = \frac{M_W}{M_Z} = \frac{|g|}{\sqrt{g^2 + g'^2}}$, and leave a massless U(1) photon, γ.

The quarks and the leptons interact with the Higgs field through Yukawa interaction terms:

where $(d, u, e, \nu)^i_{L,R}$ are left-handed and right-handed quarks and leptons of the ith generation, $\lambda^{ij}_{u,d,e}$ are matrices of Yukawa couplings where h.c. denotes the hermitian conjugate terms. In the symmetry breaking ground state, only the terms containing ϕ^0 remain, giving rise to mass terms for the fermions. Rotating the quark and lepton fields to the basis where the matrices of Yukawa couplings are diagonal, one gets

where the masses of the fermions are $m^i_{u,d,e} = \lambda^i_{u,d,e} v / \sqrt{2}$, and $\lambda^i_{u,d,e}$ denote the eigenvalues of the Yukawa matrices.[207]

7.8   See also

Standard Model

- Quantum gauge theory

- History of quantum field theory

- Introduction to quantum mechanics

- Noncommutative standard model and noncommutative geometry generally

- Standard Model (mathematical formulation) (and especially Standard Model fields overview and mass terms and the Higgs mechanism)

Other

- Bose–Einstein statistics

- Dalitz plot

- Higgs boson in fiction

- Quantum triviality

- ZZ diboson

- Scalar boson

- Stueckelberg action

- Tachyonic field

7.9   Notes

[1] Note that such events also occur due to other processes. Detection involves a statistically significant excess of such events at specific energies.

[2] In the Standard Model, the total decay width of a Higgs boson with a mass of 126 GeV/c^2 is predicted to be 4.21×10^{-3} GeV.[5] The mean lifetime is given by $\tau = \hbar/\Gamma$.

[3] The range of a force is inversely proportional to the mass of the particles transmitting it.[19] In the Standard Model, forces are carried by virtual particles. These particles' movement and interactions with each other are limited by the energy–time uncertainty principle. As a result, the more massive a single virtual particle is, the greater its energy, and therefore the shorter the distance it can travel. A particle's mass therefore determines the maximum distance at which it can interact with other particles and on any force it mediates. By the same token, the reverse is also true: massless and near-massless particles can carry long distance forces. *(See also: Compton wavelength and Static forces and virtual-particle exchange)* Since experiments have shown that the weak force acts over only a very short range, this implies that there must exist massive gauge bosons. And indeed, their masses have since been confirmed by measurement.

[4] It is quite common for a law of physics to hold true only if certain assumptions held true or only under certain conditions. For example, Newton's laws of motion apply only at speeds where relativistic effects are negligible; and laws related to conductivity, gases, and classical physics (as opposed to quantum mechanics) may apply only within certain ranges of size, temperature, pressure, or other conditions.

[5] Electroweak symmetry is broken by the Higgs field in its lowest energy state, called its "ground state". At high energy levels this does not happen, and the gauge bosons of the weak force would therefore be expected to be massless.

[6] By the 1960s, many had already started to see gauge theories as failing to explain particle physics because theorists had been unable to solve the mass problem or even explain how gauge theory could provide a solution. So the idea that the Standard Model – which relied on a Higgs field, not yet proved to exist – could be fundamentally incorrect. Against this, once the model was developed around 1972, no better theory existed, and its predictions and solutions were so accurate, that it became the preferred theory anyway. It then became crucial to science, to know whether it was *correct.*

[7] The success of the Higgs-based electroweak theory and Standard Model is illustrated by their predictions of the mass of two particles later detected: the W boson (predicted mass: 80.390 ± 0.018 GeV, experimental measurement: 80.387 ± 0.019 GeV), and the Z boson (predicted mass: 91.1874 ± 0.0021, experimental measurement: 91.1876 ± 0.0021 GeV). The existence of the Z boson was itself another prediction. Other accurate predictions included the weak neutral current, the gluon, and the top and charm quarks, all later proven to exist as the theory said.

[8] For example, Huffington Post/Reuters[35] and others[36][37]

[9] The bubble's effects would be expected to propagate across the universe at the speed of light from wherever it occurred. However space is vast – with even the nearest galaxy being over 2 million lightyears from us, and others being many billions of lightyears distant, so the effect of such an event would be unlikely to arise here for billions of years after first occurring.[39][40]

[10] If the Standard Model is valid, then the particles and forces we observe in our universe exist as they do, because of underlying quantum fields. Quantum fields can have states of differing stability, including 'stable', 'unstable' and 'metastable' states (the latter remain stable unless sufficiently perturbed). If a more stable vacuum state were able to arise, then existing particles and forces would no longer arise as they presently do. Different particles or forces would arise from (and be shaped by) whatever new quantum states arose. The world we know depends upon these particles and forces, so if this happened, everything around us, from subatomic particles to galaxies, and all fundamental forces, would be reconstituted into new fundamental particles and forces and structures. The universe would potentially lose all of its present structures and become inhabited by new ones (depending upon the exact states involved) based upon the same quantum fields.

[11] Goldstone's theorem only applies to gauges having manifest Lorentz covariance, a condition that took time to become questioned. But the process of quantisation requires a gauge to be fixed and at this point it becomes possible to choose a gauge such as the 'radiation' gauge which is not invariant over time, so that these problems can be avoided. According to Bernstein (1974, p.8):

> "the "radiation gauge" condition $\nabla \cdot A(x) = 0$ is clearly noncovariant, which means that if we wish to maintain transversality of the photon in all Lorentz frames, the photon field $A\mu(x)$ cannot transform like a four-vector. This is no catastrophe, since the photon *field* is not an observable, and one can readily show that the S-matrix elements, which *are* observable have covariant structures .... in gauge theories one might arrange things so that one had a symmetry breakdown because of the noninvariance of the vacuum; but, because the Goldstone *et al.* proof breaks down, the zero mass Goldstone mesons need not appear." [Emphasis in original]

Bernstein (1974) contains an accessible and comprehensive background and review of this area, see external links

[12] A field with the "Mexican hat" potential $V(\phi) = \mu^2\phi^2 + \lambda\phi^4$ and $\mu^2 < 0$ has a minimum not at zero but at some non-zero value ϕ_0. By expressing the action in terms of the field $\tilde{\phi} = \phi - \phi_0$ (where ϕ_0 is a constant independent of position), we find the Yukawa term has a component $g\phi_0\bar{\psi}\psi$. Since both g and ϕ_0 are constants, this looks exactly like the mass term for a fermion of mass $g\phi_0$. The field $\tilde{\phi}$ is then the Higgs field.

[13] In the Standard Model, the mass term arising from the Dirac Lagrangian for any fermion ψ is $-m\bar{\psi}\psi$. This is *not* invariant under the electroweak symmetry, as can be seen by writing ψ in terms of left and right handed components:

$$-m\bar{\psi}\psi = -m(\bar{\psi}_L\psi_R + \bar{\psi}_R\psi_L)$$

i.e., contributions from $\bar{\psi}_L\psi_L$ and $\bar{\psi}_R\psi_R$ terms do not appear. We see that the mass-generating interaction is achieved by constant flipping of particle chirality. Since the spin-half particles have no right/left helicity pair with the same SU(2) and SU(3) representation and the same weak hypercharge, then assuming these gauge charges are conserved in the vacuum, none of the spin-half particles could ever swap helicity. Therefore, in the absence of some other cause, all fermions must be massless.

[14] The example is based on the production rate at the LHC operating at 7 TeV. The total cross-section for producing a Higgs boson at the LHC is about 10 picobarn,[97] while the total cross-section for a proton–proton collision is 110 millibarn.[100]

[15] Just before LEP's shut down, some events that hinted at a Higgs were observed, but it was not judged significant enough to extend its run and delay construction of the LHC.

[16] Announced in articles in *Time*,[145] Forbes,[146] *Slate*,[147] NPR,[148] and others.[149]

[17] Other names have included: the "Anderson–Higgs" mechanism,[159] "Higgs–Kibble" mechanism (by Abdus Salam)[79] and "ABEGHHK'tH" mechanism [for Anderson, Brout, Englert, Guralnik, Hagen, Higgs, Kibble and 't Hooft] (by Peter Higgs).[79]

[18] In Miller's analogy, the Higgs field is compared to political party workers spread evenly throughout a room. There will be some people (in Miller's example an anonymous person) who pass through the crowd with ease, paralleling the interaction between the field and particles that do not interact with it, such as massless photons. There will be other people (in Miller's example the British prime minister) who would find their progress being continually slowed by the swarm of admirers crowding around, paralleling the interaction for particles that do interact with the field and by doing so, acquire a finite mass.[198][199]

7.10   References

[1] O'Luanaigh, C. (14 March 2013). "New results indicate that new particle is a Higgs boson". CERN. Retrieved 2013-10-09.

[2] Bryner, J. (14 March 2013). "Particle confirmed as Higgs boson". *NBC News*. Retrieved 2013-03-14.

[3] Heilprin, J. (14 March 2013). "Higgs Boson Discovery Confirmed After Physicists Review Large Hadron Collider Data at CERN". *The Huffington Post*. Retrieved 2013-03-14.

[4] ATLAS; CMS (26 March 2015). "Combined Measurement of the Higgs Boson Mass in pp Collisions at √s=7 and 8 TeV with the ATLAS and CMS Experiments". arXiv:1503.07589.

[5] LHC Higgs Cross Section Working Group; Dittmaier; Mariotti; Passarino; Tanaka; Alekhin; Alwall; Bagnaschi; Banfi (2012). "Handbook of LHC Higgs Cross Sections: 2. Differential Distributions". *CERN Report 2 (Tables A.1 – A.20)* **1201**: 3084. arXiv:1201.3084. Bibcode:2012arXiv1201.3084L.

[6] Onyisi, P. (23 October 2012). "Higgs boson FAQ". University of Texas ATLAS group. Retrieved 2013-01-08.

[7] Strassler, M. (12 October 2012). "The Higgs FAQ 2.0". *ProfMattStrassler.com*. Retrieved 2013-01-08. [Q] Why do particle physicists care so much about the Higgs particle?
[A] Well, actually, they don't. What they really care about is the Higgs *field*, because it is *so* important. [emphasis in original]

[8] José Luis Lucio and Arnulfo Zepeda (1987). *Proceedings of the II Mexican School of Particles and Fields, Cuernavaca-Morelos, 1986*. World Scientific. p. 29. ISBN 9971504340.

[9] Gunion, Dawson, Kane, and Haber (199). *The Higgs Hunter's Guide (1st ed.)*. pp. 11 (?). ISBN 9780786743186. – quoted as being in the first (1990) edition of the book by Peter Higgs in his talk "My Life as a Boson", 2001, ref#25.

[10] Strassler, M. (8 October 2011). "The Known Particles – If The Higgs Field Were Zero". *ProfMattStrassler.com*. Retrieved 13 November 2012. The Higgs field: so important it merited an entire experimental facility, the Large Hadron Collider, dedicated to understanding it.

[11] Biever, C. (6 July 2012). "It's a boson! But we need to know if it's the Higgs". *New Scientist*. Retrieved 2013-01-09. 'As a layman, I would say, I think we have it,' said Rolf-Dieter Heuer, director general of CERN at Wednesday's seminar announcing the results of the search for the Higgs boson. But when pressed by journalists afterwards on what exactly 'it' was, things got more complicated. 'We have discovered a boson – now we have to find out what boson it is'
Q: 'If we don't know the new particle is a Higgs, what do we know about it?' We know it is some kind of boson, says Vivek Sharma of CMS [...]
Q: 'are the CERN scientists just being too cautious? What would be enough evidence to call it a Higgs boson?' As there could be many different kinds of Higgs bosons, there's no straight answer.
[emphasis in original]

[12] Siegfried, T. (20 July 2012). "Higgs Hysteria". *Science News*. Retrieved 2012-12-09. In terms usually reserved for athletic achievements, news reports described the finding as a monumental milestone in the history of science.

[13] Del Rosso, A. (19 November 2012). "Higgs: The beginning of the exploration". *CERN Bulletin* (47–48). Retrieved 2013-01-09. Even in the most specialized circles, the new particle discovered in July is not yet being called the "Higgs boson". Physicists still hesitate to call it that before they have determined that its properties fit with those the Higgs theory predicts the Higgs boson has.

[14] Naik, G. (14 March 2013). "New Data Boosts Case for Higgs Boson Find". *The Wall Street Journal*. Retrieved 2013-03-15. 'We've never seen an elementary particle with spin zero,' said Tony Weidberg, a particle physicist at the University of Oxford who is also involved in the CERN experiments.

[15] Overbye, D. (8 October 2013). "For Nobel, They Can Thank the 'God Particle'". *The New York Times*. Retrieved 2013-11-03.

[16] Sample, I. (29 May 2009). "Anything but the God particle". *The Guardian*. Retrieved 2009-06-24.

[17] Evans, R. (14 December 2011). "The Higgs boson: Why scientists hate that you call it the 'God particle'". *National Post*. Retrieved 2013-11-03.

[18] The nickname occasionally has been satirised in mainstream media as well. Borowitz, Andy (July 13, 2012). "5 questions for the Higgs boson". *The New Yorker*.

[19] Shu, F. H. (1982). *The Physical Universe: An Introduction to Astronomy*. University Science Books. pp. 107–108. ISBN 978-0-935702-05-7.

[20] Shu, F. H. (1982). *The Physical Universe: An Introduction to Astronomy*. University Science Books. pp. 107–108. ISBN 978-0-935702-05-7.

[21] Leon M. Lederman and Dick Teresi (1993). *The God Particle: If the Universe is the Answer, What is the Question*. Houghton Mifflin Company.

[22] Heath, Nick, *The Cern tech that helped track down the God particle*, TechRepublic, 4 July 2012

[23] Rao, Achintya (2 July 2012). "Why would I care about the Higgs boson?". *CMS Public Website*. CERN. Retrieved 18 July 2012.

[24] Max Jammer, *Concepts of Mass in Contemporary Physics and Philosophy* (Princeton, NJ: Princeton University Press, 2000) pp.162–163, who provides many references in support of this statement.

[25] The Large Hadron Collider: Shedding Light on the Early Universe – lecture by R.-D. Heuer, CERN, Chios, Greece, 28 September 2011

[26] Alekhin, Djouadi and Moch, S.; Djouadi, A.; Moch, S. (2012-08-13). "The top quark and Higgs boson masses and the stability of the electroweak vacuum". *Physics Letters B* **716**: 214. arXiv:1207.0980. Bibcode:2012PhLB..716..214A. doi:10.1016/j. 24.Retrieved20February2013.

[27] M.S. Turner, F. Wilczek (1982). "Is our vacuum metastable?". *Nature* **298** (5875): 633–634. Bibcode:1982Natur.298..633T. doi:10.1038/298633a0.

[28] S. Coleman and F. De Luccia (1980). "Gravitational effects on and of vacuum decay". *Physical Review* **D21** (12): 3305. Bibcode:1980PhRvD..21.3305C. doi:10.1103/PhysRevD.21.3305.

[29] M.Stone(1976). "Lifetime and decay of excited vacuum states".*Phys.Rev.D***14**(12): 3568–3573. Bibcode:1976PhRvD..14.35. doi:10.1103/PhysRevD.14.3568.

[30] P.H.Frampton(1976). "Vacuum Instability and Higgs Scalar Mass".*Phys.Rev.Lett.***37**(21): 1378–1380. Bibcode:1976Ph1378F. doi:10.1103/PhysRevLett.37.1378.

[31] P.H. Frampton (1977). "Consequences of Vacuum Instability in Quantum Field Theory". *Phys. Rev.* **D15** (10): 2922–28. Bibcode:1977PhRvD..15.2922F. doi:10.1103/PhysRevD.15.2922.

[32] Ellis, Espinosa, Giudice, Hoecker, & Riotto, J.; Espinosa, J.R.; Giudice, G.F.; Hoecker, A.; Riotto, A. (2009). "The Probable Fate of the Standard Model". *Phys. Lett. B* **679** (4): 369–375. arXiv:0906.0954. Bibcode:2009PhLB..679..369E. doi:10.1016/j.physletb.2009.07.054.

[33] Masina, Isabella (2013-02-12). "Higgs boson and top quark masses as tests of electroweak vacuum stability". *Phys. Rev. D* **87** (5): 53001. arXiv:1209.0393. Bibcode:2013PhRvD..87e3001M. doi:10.1103/PhysRevD.87.053001.

[34] Buttazzo, Degrassi, Giardino, Giudice, Sala, Salvio, Strumia (2013-07-12). "Investigating the near-criticality of the Higgs boson". *JHEP 1312 (2013) 089*. arXiv:1307.3536. Bibcode:2013JHEP...12..089B. doi:10.1007/JHEP12(2013)089.

[35] Irene Klotz (editing by David Adams and Todd Eastham) (2013-02-18). "Universe Has Finite Lifespan, Higgs Boson Calculations Suggest". *Huffington Post*. Reuters. Retrieved 21 February 2013. Earth will likely be long gone before any Higgs boson particles set off an apocalyptic assault on the universe

[36] Hoffman, Mark (2013-02-19). "Higgs Boson Will Destroy The Universe Eventually". *ScienceWorldReport*. Retrieved 21 February 2013.

[37] "Higgs boson will aid in creation of the universe – and how it will end". *Catholic Online/NEWS CONSORTIUM*. 2013-02-20. Retrieved 21 February 2013. [T]he Earth will likely be long gone before any Higgs boson particles set off an apocalyptic assault on the universe

[38] Salvio, Alberto (2015-04-09). "A Simple Motivated Completion of the Standard Model below the Planck Scale: Axions and Right-Handed Neutrinos".*Physics Letters B***743**: 428. arXiv:1501.03781. Bibcode:2015PhLB..743..428S.doi:10.1016/j.phys5.

[39] Boyle, Alan (2013-02-19). "Will our universe end in a 'big slurp'? Higgs-like particle suggests it might". *NBC News' Cosmic log*. Retrieved 21 February 2013. [T]he bad news is that its mass suggests the universe will end in a fast-spreading bubble of doom. The good news? It'll probably be tens of billions of years. The article quotes Fermilab's Joseph Lykken: "[T]he parameters for our universe, including the Higgs [and top quark's masses] suggest that we're just at the edge of stability, in a "metastable" state. Physicists have been contemplating such a possibility for more than 30 years. Back in 1982, physicists Michael Turner and Frank Wilczek wrote in Nature that "without warning, a bubble of true vacuum could nucleate somewhere in the universe and move outwards..."

[40] Peralta, Eyder (2013-02-19). "If Higgs Boson Calculations Are Right, A Catastrophic 'Bubble' Could End Universe". *npr – two way*. Retrieved 21 February 2013. Article cites Fermilab's Joseph Lykken: "The bubble forms through an unlikely quantum fluctuation, at a random time and place," Lykken tells us. "So in principle it could happen tomorrow, but then most likely in a very distant galaxy, so we are still safe for billions of years before it gets to us."

[41] Bezrukov; Shaposhnikov (2007-10-19). "The Standard Model Higgs boson as the inflaton". *Phys.Lett. B659 (2008) 703-706*. arXiv:0710.3755. Bibcode:2008PhLB..659..703B. doi:10.1016/j.physletb.2007.11.072.

[42] Salvio, Alberto (2013-08-09). "Higgs Inflation at NNLO after the Boson Discovery". *Phys.Lett. B727 (2013) 234-239*. arXiv:1308.2244. Bibcode:2013PhLB..727..234S. doi:10.1016/j.physletb.2013.10.042.

[43] Cole, K. (2000-12-14). "One Thing Is Perfectly Clear: Nothingness Is Perfect". *Los Angeles Times*. p. 'Science File'. Retrieved 17 January 2013. [T]he Higgs' influence (or the influence of something like it) could reach much further. For example, something like the Higgs—if not exactly the Higgs itself—may be behind many other unexplained "broken symmetries" in the universe as well ... In fact, something very much like the Higgs may have been behind the collapse of the symmetry that led to the Big Bang, which created the universe. When the forces first began to separate from their primordial sameness—taking on the distinct characters they have today—they released energy in the same way as water releases energy when it turns to ice. Except in this case, the freezing packed enough energy to blow up the universe. ... However it happened, the moral is clear: Only when the perfection shatters can everything else be born.

[44] Higgs Matters – Kathy Sykes, 30 Nove 2012

[45] Why the public should care about the Higgs Boson – Jodi Lieberman, American Physical Society (APS)

[46] Matt Strassler's blog – Why the Higgs particle matters 2 July 2012

[47] Sean Carroll (13 November 2012). *The Particle at the End of the Universe: How the Hunt for the Higgs Boson Leads Us to the Edge of a New World*. Penguin Group US. ISBN 978-1-101-60970-5.

[48] Woit, Peter (13 November 2010). "The Anderson–Higgs Mechanism". Dr. Peter Woit (Senior Lecturer in Mathematics Columbia University and Ph.D. particle physics). Retrieved 12 November 2012.

[49] Goldstone,J;Salam,Abdus;Weinberg,Steven(1962). "Broken Symmetries".*Physical Review***127**(3): 965–970. Bibcode:1965G. doi:10.1103/PhysRev.127.965.

[50] Guralnik, G. S. (2011). "The Beginnings of Spontaneous Symmetry Breaking in Particle Physics". arXiv:1110.2253 [physics.hist-ph].

[51] Kibble,T.W.B. (2009). "Englert–Brout–Higgs–Guralnik–Hagen–Kibble Mechanism".*Scholarpedia***4**(1): 6441. Bibc441K. doi:10.4249/scholarpedia.6441. Retrieved 2012-11-23.

[52] Kibble, T. W. B. "History of Englert–Brout–Higgs–Guralnik–Hagen–Kibble Mechanism (history)". *Scholarpedia* **4** (1): 8741. Bibcode:2009SchpJ...4.8741K. doi:10.4249/scholarpedia.8741. Retrieved 2012-11-23.

[53] The Nobel Prize in Physics 2008 – official Nobel Prize website.

[54] List of Anderson 1958–1959 papers referencing 'symmetry', at APS Journals

[55] Higgs, Peter (2010-11-24). "My Life as a Boson" (PDF). Talk given by Peter Higgs at Kings College, London, Nov 24 2010, expanding on a paper originally presented in 2001. Retrieved 17 January 2013. – the original 2001 paper can be found at: Duff and Liu, ed. (2003) [year of publication]. *2001 A Spacetime Odyssey: Proceedings of the Inaugural Conference of the Michigan Center for Theoretical Physics, Michigan, USA, 21–25 May 2001.* World Scientific. pp. 86–88. ISBN 9812382313. Retrieved 17 January 2013.

[56] Anderson, P. (1963). "Plasmons, gauge invariance and mass". *Physical Review* **130**: 439. Bibcode:1963PhRv..130..439A. doi:10.1103/PhysRev.130.439.

[57] Klein, A.; Lee, B. (1964). "Does Spontaneous Breakdown of Symmetry Imply Zero-Mass Particles?". *Physical Review Letters* **12** (10): 266. Bibcode:1964PhRvL..12..266K. doi:10.1103/PhysRevLett.12.266.

[58] Englert, François; Brout, Robert (1964). "Broken Symmetry and the Mass of Gauge Vector Mesons". *Physical Review Letters* **13** (9): 321–23. Bibcode:1964PhRvL..13..321E. doi:10.1103/PhysRevLett.13.321.

[59] Higgs, Peter (1964). "Broken Symmetries and the Masses of Gauge Bosons". *Physical Review Letters* **13** (16): 508–509. Bibcode:1964PhRvL..13..508H. doi:10.1103/PhysRevLett.13.508.

[60] Guralnik, Gerald; Hagen, C. R.; Kibble, T. W. B. (1964). "Global Conservation Laws and Massless Particles". *Physical Review Letters* **13** (20): 585–587. Bibcode:1964PhRvL..13..585G. doi:10.1103/PhysRevLett.13.585.

[61] Higgs,Peter(1964). "Broken symmetries,massless particles and gaugefields".*Physics Letters***12**(2): 132–133. Bibcode:196432H. doi:10.1016/0031-9163(64)91136-9.

[62] Higgs, Peter (2010-11-24). "My Life as a Boson" (PDF). Talk given by Peter Higgs at Kings College, London, Nov 24 2010. Retrieved 17 January 2013. Gilbert ... wrote a response to [Klein and Lee's paper] saying 'No, you cannot do that in a relativistic theory. You cannot have a preferred unit time-like vector like that.' This is where I came in, because the next month was when I responded to Gilbert's paper by saying 'Yes, you can have such a thing' but only in a gauge theory with a gauge field coupled to the current.

[63] G.S. Guralnik (2011). "Gauge invariance and the Goldstone theorem – 1965 Feldafing talk". *Modern Physics Letters A* **26** (19): 1381–1392. arXiv:1107.4592. Bibcode:2011MPLA...26.1381G. doi:10.1142/S0217732311036188.

[64] Higgs, Peter (1966). "Spontaneous Symmetry Breakdown without Massless Bosons". *Physical Review* **145** (4): 1156–1163. Bibcode:1966PhRv..145.1156H. doi:10.1103/PhysRev.145.1156.

[65] Kibble,Tom(1967). "Symmetry Breaking in Non-Abelian Gauge Theories".*Physical Review***155**(5): 1554–1561. BibcodK. doi:10.1103/PhysRev.155.1554.

[66] "Guralnik, G S; Hagen, C R and Kibble, T W B (1967). Broken Symmetries and the Goldstone Theorem. Advances in Physics, vol. 2" (PDF).

[67] "Physical Review Letters – 50th Anniversary Milestone Papers". Physical Review Letters.

[68] S. Weinberg (1967). "A Model of Leptons". *Physical Review Letters* **19** (21): 1264–1266. Bibcode:1967PhRvL..19.1264W. doi:10.1103/PhysRevLett.19.1264.

[69] A. Salam (1968). N. Svartholm, ed. *Elementary Particle Physics: Relativistic Groups and Analyticity.* Eighth Nobel Symposium. Stockholm: Almquvist and Wiksell. p. 367.

[70] S.L.Glashow(1961). "Partial-symmetries of weak interactions".*Nuclear Physics***22**(4): 579–588. Bibcode:1961NucPh..22..5G. doi:10.1016/0029-5582(61)90469-2.

[71] Ellis, John; Gaillard, Mary K.; Nanopoulos, Dimitri V. (2012). "A Historical Profile of the Higgs Boson". arXiv:1201.6045 [hep-ph].

[72] "Martin Veltman Nobel Lecture, December 12, 1999, p.391" (PDF). Retrieved 2013-10-09.

[73] Politzer, David. "The Dilemma of Attribution". *Nobel Prize lecture, 2004.* Nobel Prize. Retrieved 22 January 2013. Sidney Coleman published in Science magazine in 1979 a citation search he did documenting that essentially no one paid any attention to Weinberg's Nobel Prize winning paper until the work of 't Hooft (as explicated by Ben Lee). In 1971 interest in Weinberg's paper exploded. I had a parallel personal experience: I took a one-year course on weak interactions from Shelly Glashow in 1970, and he never even mentioned the Weinberg–Salam model or his own contributions.

[74] Coleman,Sidney(1979-12-14). "The1979Nobel Prize in Physics".*Science***206**(4424): 1290–1292. Bibcode:1979Sci...206.C. doi:10.1126/science.206.4424.1290.

[75] Letters from the Past – A PRL Retrospective (50 year celebration, 2008)

[76] Jeremy Bernstein (January 1974). "Spontaneous symmetry breaking, gauge theories, the Higgs mechanism and all that" (PDF). *Reviews of Modern Physics* **46** (1): 7. Bibcode:1974RvMP...46....7B. doi:10.1103/RevModPhys.46.7. Retrieved 2012-12-10.

[77] American Physical Society – "J. J. Sakurai Prize for Theoretical Particle Physics".

[78] Merali, Zeeya (4 August 2010). "Physicists get political over Higgs". *Nature Magazine*. Retrieved 28 December 2011.

[79] Close, Frank (2011). *The Infinity Puzzle: Quantum Field Theory and the Hunt for an Orderly Universe*. Oxford: Oxford University Press. ISBN 978-0-19-959350-7.

[80] G.S. Guralnik (2009). "The History of the Guralnik, Hagen and Kibble development of the Theory of Spontaneous Symmetry Breaking and Gauge Particles". *International Journal of Modern Physics A* **24** (14): 2601–2627. arXiv:0907.3466. Bibcode:2009IJMPA..24.2601G. doi:10.1142/S0217751X09045431.

[81] Peskin, Michael E.; Schroeder, Daniel V. (1995). *Introduction to Quantum Field Theory*. Reading, MA: Addison-Wesley Publishing Company. pp. 717–719 and 787–791. ISBN 0-201-50397-2.

[82] Peskin & Schroeder 1995, pp. 715–716

[83] Gunion, John (2000). *The Higgs Hunter's Guide* (illustrated, reprint ed.). Westview Press. pp. 1–3. ISBN 9780738203058.

[84] Lisa Randall, *Warped Passages: Unraveling the Mysteries of the Universe's Hidden Dimensions*, p.286: "People initially thought of tachyons as particles travelling faster than the speed of light...But we now know that a tachyon indicates an instability in a theory that contains it. Regrettably for science fiction fans, tachyons are not real physical particles that appear in nature."

[85] Sen,Ashoke(April2002). "Rolling Tachyon".*J.High Energy Phys.***2002**(0204): 048. arXiv:hep-th/0203211. Bibcode:2002JH48S. doi:10.1088/1126-6708/2002/04/048.

[86] Kutasov, David; Marino, Marcos & Moore, Gregory W. (2000). "Some exact results on tachyon condensation in string field theory". *JHEP* **0010**: 045.

[87] Aharonov, Y.; Komar, A.; Susskind, L. (1969). "Superluminal Behavior, Causality, and Instability". *Phys. Rev.* (American Physical Society) **182** (5): 1400–1403. Bibcode:1969PhRv..182.1400A. doi:10.1103/PhysRev.182.1400.

[88] Feinberg,Gerald(1967). "Possibility of Faster-Than-Light Particles".*Physical Review***159**(5): 1089–1105. Bibcode:1967PhRv. doi:10.1103/PhysRev.159.1089.

[89] Michael E. Peskin and Daniel V. Schroeder (1995). *An Introduction to Quantum Field Theory*, Perseus books publishing.

[90] Flatow, Ira (6 July 2012). "At Long Last, The Higgs Particle... Maybe". *NPR*. Retrieved 10 July 2012.

[91] "Explanatory Figures for the Higgs Boson Exclusion Plots". *ATLAS News*. CERN. Retrieved 6 July 2012.

[92] Bernardi, G.; Carena, M.; Junk, T. (2012). "Higgs Bosons: Theory and Searches" (PDF). p. 7.

[93] Lykken, Joseph D. (2009). "Beyond the Standard Model". *Proceedings of the 2009 European School of High-Energy Physics, Bautzen, Germany, 14 – 27 June 2009*. arXiv:1005.1676.

[94] Plehn, Tilman (2012). *Lectures on LHC Physics*. Lecture Notes is Physics **844**. Springer. Sec. 1.2.2. arXiv:0910.4122. ISBN 3642240399.

[95] "LEP Electroweak Working Group".

[96] Peskin, Michael E.; Wells, James D. (2001). "How Can a Heavy Higgs Boson be Consistent with the Precision Electroweak Measurements?".*Physical Review D***64**(9): 093003. arXiv:hep-ph/0101342. Bibcode:2001PhRvD..64i3003P.doi:10.1103/003.

[97] Baglio, Julien; Djouadi, Abdelhak (2011). "Higgs production at the lHC". *Journal of High Energy Physics* **1103** (3): 055. arXiv:1012.0530. Bibcode:2011JHEP...03..055B. doi:10.1007/JHEP03(2011)055.

[98] Baglio, Julien; Djouadi, Abdelhak (2010). "Predictions for Higgs production at the Tevatron and the associated uncertainties". *Journal of High Energy Physics***1010**(10): 063. arXiv:1003.4266. Bibcode:2010JHEP...10..064B.doi:10.1007/JHEP10(4.

[99] Teixeira-Dias (LEP Higgs working group), P. (2008). "Higgs boson searches at LEP". *Journal of.Physics: Conference Series* **110** (4): 042030. arXiv:0804.4146. Bibcode:2008JPhCS.110d2030T. doi:10.1088/1742-6596/110/4/042030.

[100] "Collisions". *LHC Machine Outreach*. CERN. Retrieved 26 July 2012.

[101] Asquith, Lily (22 June 2012). "Why does the Higgs decay?". *Life and Physics* (London: The Guardian). Retrieved 14 August 2012.

[102] "Higgs bosons: theory and searches" (PDF). *PDGLive*. Particle Data Group. 12 July 2012. Retrieved 15 August 2012.

[103] Branco, G. C.; Ferreira, P.M.; Lavoura, L.; Rebelo, M.N.; Sher, Marc; Silva, João P. (July 2012). "Theory and phenomenology of two-Higgs-doublet models". *Physics Reports* (Elsevier) **516** (1): 1–102. arXiv:1106.0034. Bibcode:2012PhR...516....1B. doi:10.1016/j.physrep.2012.02.002.

[104] Csaki, C.; Grojean, C.; Pilo, L.; Terning, J. (2004). "Towards a realistic model of Higgsless electroweak symmetry breaking". *Physical Review Letters* **92** (10): 101802. arXiv:hep-ph/0308038. Bibcode:2004PhRvL..92j1802C. doi:10.1103/PhysRevLe .PMID15089195.

[105] Csaki, C.; Grojean, C.; Pilo, L.; Terning, J.; Terning, John (2004). "Gauge theories on an interval: Unitarity without a Higgs". *Physical Review D* **69** (5): 055006. arXiv:hep-ph/0305237. Bibcode:2004PhRvD..69e5006C. doi:10.1103/PhysRevD.69.055006.

[106] "The Hierarchy Problem: why the Higgs has a snowball's chance in hell". Quantum Diaries. 2012-07-01. Retrieved 19 March 2013.

[107] "The Hierarchy Problem | Of Particular Significance". Profmattstrassler.com. Retrieved 2013-10-09.

[108] "Collisions". *LHC Machine Outreach*. CERN. Retrieved 26 July 2012.

[109] "Hunt for Higgs boson hits key decision point". MSNBC. 2012-12-06. Retrieved 2013-01-19.

[110] Worldwide LHC Computing Grid main page 14 November 2012: *"[A] global collaboration of more than 170 computing centres in 36 countries ... to store, distribute and analyse the ~25 Petabytes (25 million Gigabytes) of data annually generated by the Large Hadron Collider"*

[111] What is the Worldwide LHC Computing Grid? (Public 'About' page) 14 November 2012: *"Currently WLCG is made up of more than 170 computing centers in 36 countries...The WLCG is now the world's largest computing grid"*

[112] W.-M.Yao;et al. (2006). "Review of Particle Physics"(PDF).*Journal of Physics G* **33**: 1. arXiv:astro-ph/0601168. BibcodeY. doi:10,1088/0954-3899/33/1/001.

[113] The CDF Collaboration, the D0 Collaboration, the Tevatron New Physics, Higgs Working Group (2012). "Updated Combination of CDF and D0 Searches for Standard Model Higgs Boson Production with up to 10.0 fb^{-1} of Data". arXiv:1207.0449 [hep-ex].

[114] "Interim Summary Report on the Analysis of the 19 September 2008 Incident at the LHC" (PDF). CERN. 15 October 2008. EDMS 973073. Retrieved 28 September 2009.

[115] "CERN releases analysis of LHC incident" (Press release). CERN Press Office. 16 October 2008. Retrieved 28 September 2009.

[116] "LHC to restart in 2009" (Press release). CERN Press Office. 5 December 2008. Retrieved 8 December 2008.

[117] "LHC progress report". *The Bulletin*. CERN. 3 May 2010. Retrieved 7 December 2011.

[118] "ATLAS experiment presents latest Higgs search status". *ATLAS homepage*. CERN. 13 December 2011. Retrieved 13 December 2011.

[119] Taylor, Lucas (13 December 2011). "CMS search for the Standard Model Higgs Boson in LHC data from 2010 and 2011". *CMS public website*. CERN. Retrieved 13 December 2011.

[120] Overbye, D. (5 March 2013). "Chasing The Higgs Boson". *The New York Times*. Retrieved 2013-03-05.

[121] "ATLAS and CMS experiments present Higgs search status" (Press release). CERN Press Office. 13 December 2011. Retrieved 14 September 2012. the statistical significance is not large enough to say anything conclusive. As of today what we see is consistent either with a background fluctuation or with the presence of the boson. Refined analyses and additional data delivered in 2012 by this magnificent machine will definitely give an answer

[122] "WLCG Public Website". CERN. Retrieved 29 October 2012.

[123] CMS collaboration (2014). "Precise determination of the mass of the Higgs boson and tests of compatibility of its couplings with the standard model predictions using proton collisions at 7 and 8 TeV". arXiv:1412.8662.

[124] ATLAS collaboration (2014). "Measurements of Higgs boson production and couplings in the four-lepton channel in pp collisions at center-of-mass energies of 7 and 8 TeV with the ATLAS detector". arXiv:1408.5191.

[125] ATLAS collaboration (2014). "Measurement of Higgs boson production in the diphoton decay channel in pp collisions at center-of-mass energies of 7 and 8 TeV with the ATLAS detector". arXiv:1408.7084.

[126] "Press Conference: Update on the search for the Higgs boson at CERN on 4 July 2012". Indico.cern.ch. 22 June 2012. Retrieved 4 July 2012.

[127] "CERN to give update on Higgs search". CERN. 22 June 2012. Retrieved 2 July 2011.

[128] "Scientists analyse global Twitter gossip around Higgs boson discovery". *phys.org (from arXiv)*. 2013-01-23. Retrieved 6 February 2013. – stated to be *" the first time scientists have been able to analyse the dynamics of social media on a global scale before, during and after the announcement of a major scientific discovery."* For the paper itself see: De Domenico, M.; Lima, A.; Mougel, P.; Musolesi, M. (2013). "The Anatomy of a Scientific Gossip". arXiv:1301.2952. Bibcode:2013NatSR...3E2980D. doi:10.1038/srep02980.

[129] "Higgs boson particle results could be a quantum leap". Times LIVE. 28 June 2012. Retrieved 4 July 2012.

[130] CERN prepares to deliver Higgs particle findings, Australian Broadcasting Corporation. Retrieved 4 July 2012.

[131] "God Particle Finally Discovered? Higgs Boson News At Cern Will Even Feature Scientist It's Named After". Huffingtonpost.co.uk. Retrieved 2013-01-19.

[132] Our Bureau (2012-07-04). "Higgs on way, theories thicken". Calcutta, India: Telegraphindia.com. Retrieved 2013-01-19.

[133] Thornhill, Ted (2013-07-03). "God Particle Finally Discovered? Higgs Boson News At Cern Will Even Feature Scientist It's Named After". *Huffington Post*. Retrieved 23 July 2013.

[134] Cooper, Rob (2013-07-01) [updated subsequently]. "God particle is 'found': Scientists at Cern expected to announce on Wednesday Higgs boson particle has been discovered". *Daily Mail* (London). Retrieved 23 July 2013. - States that *"Five leading theoretical physicists have been invited to the event on Wednesday - sparking speculation that the particle has been discovered."*, including Higgs and Englert, and that Kibble - who was invited but unable to attend - "told the Sunday Times: 'My guess is that is must be a pretty positive result for them to be asking us out there'."

[135] Adrian Cho (13 July 2012). "Higgs Boson Makes Its Debut After Decades-Long Search". *Science* **337** (6091): 141–143. doi:10.1126/science.337.6091.141. PMID 22798574.

[136] CMS collaboration (2012). "Observation of a new boson at a mass of 125 GeV with the CMS experiment at the LHC". *Physics Letters B* **716** (1): 30–61. arXiv:1207.7235. Bibcode:2012PhLB..716...30C. doi:10.1016/j.physletb.2012.08.021.

[137] Taylor, Lucas (4 July 2012). "Observation of a New Particle with a Mass of 125 GeV". *CMS Public Website*. CERN. Retrieved 4 July 2012.

[138] "Latest Results from ATLAS Higgs Search". *ATLAS News*. CERN. 4 July 2012. Retrieved 4 July 2012.

[139] ATLAS collaboration (2012). "Observation of a New Particle in the Search for the Standard Model Higgs Boson with the ATLAS Detector at the LHC". *Physics Letters B* **716**(1): 1–29. arXiv:1207.7214. Bibcode:2012PhLB..716....1A. doi:10.1016/j.0.

[140] Gillies, James (23 July 2012). "LHC 2012 proton run extended by seven weeks". *CERN bulletin*. Retrieved 29 August 2012.

[141] "Higgs boson behaving as expected". *3 News NZ*. 15 November 2012.

[142] Strassler, Matt (2012-11-14). "Higgs Results at Kyoto". *Of Particular Significance: Conversations About Science with Theoretical Physicist Matt Strassler*. Prof. Matt Strassler's personal particle physics website. Retrieved 10 January 2013. ATLAS and CMS only just co-discovered this particle in July ... We will not know after today whether it is a Higgs at all, whether it is a Standard Model Higgs or not, or whether any particular speculative idea...is now excluded. [...] Knowledge about nature does not come easy. We discovered the top quark in 1995, and we are still learning about its properties today... we will still be learning important things about the Higgs during the coming few decades. We've no choice but to be patient.

[143] Sample, Ian (14 November 2012). "Higgs particle looks like a bog Standard Model boson, say scientists". *The Guardian* (London). Retrieved 15 November 2012.

[144] "CERN experiments observe particle consistent with long-sought Higgs boson". CERN press release. 4 July 2012. Retrieved 4 July 2012.

[145] "Person Of The Year 2012". *Time*. 19 December 2012.

[146] "Higgs Boson Discovery Has Been Confirmed". Forbes. Retrieved 2013-10-09.

[147] Slate Video Staff (2012-09-11). "Higgs Boson Confirmed; CERN Discovery Passes Test". Slate.com. Retrieved 2013-10-09.

[148] "The Year Of The Higgs, And Other Tiny Advances In Science". NPR. 2013-01-01. Retrieved 2013-10-09.

[149] "Confirmed: the Higgs boson does exist". *The Sydney Morning Herald*. 4 July 2012.

[150] "AP CERN chief: Higgs boson quest could wrap up by midyear". *MSNBC*. Associated Press. 2013-01-27. Retrieved 20 February 2013. Rolf Heuer, director of [CERN], said he is confident that "towards the middle of the year, we will be there." – Interview by AP, at the World Economic Forum, 26 Jan 2013.

[151] Boyle, Alan (2013-02-16). "Will our universe end in a 'big slurp'? Higgs-like particle suggests it might". *NBCNews.com – cosmic log*. Retrieved 20 February 2013. 'it's going to take another few years' after the collider is restarted to confirm definitively that the newfound particle is the Higgs boson.

[152] Gillies, James (2013-03-06). "A question of spin for the new boson". CERN. Retrieved 7 March 2013.

[153] Adam Falkowski (writing as 'Jester') (2013-02-27). "When shall we call it Higgs?". Résonaances particle physics blog. Retrieved 7 March 2013.

[154] CMS Collaboration (February 2013). "Study of the Mass and Spin-Parity of the Higgs Boson Candidate via Its Decays to Z Boson Pairs". *Phys. Rev. Lett.* (American Physical Society) **110** (8): 081803. arXiv:1212.6639. Bibcode:2013PhRvL.110h1803C. doi:10.1103/PhysRevLett.110.081803. Retrieved 15 September 2014.

[155] ATLAS Collaboration (7 October 2013). "Evidence for the spin-0 nature of the Higgs boson using ATLAS data". *Phys. Lett. B* (American Physical Society) **726** (1-3): 120–144. Bibcode:2013PhLB..726..120A. doi:10.1016/j.physletb.2013.08.026. Retrieved 15 September 2014.

[156] "Higgs-like Particle in a Mirror". American Physical Society. Retrieved 26 February 2013.

[157] The CMS Collaboration (2014-06-22). "Evidence for the direct decay of the 125 GeV Higgs boson to fermions". Nature Publishing Group doi= 10.1038/nphys3005.

[158] Adam Falkowski (writing as 'Jester') (2012-12-13). "Twin Peaks in ATLAS". Résonaances particle physics blog. Retrieved 24 February 2013.

[159] Liu, G. Z.; Cheng, G. (2002). "Extension of the Anderson-Higgs mechanism". *Physical Review B* **65** (13): 132513. arXiv:cond-mat/0106070. Bibcode:2002PhRvB..65m2513L. doi:10.1103/PhysRevB.65.132513.

[160] Editorial (2012-03-21). "Mass appeal: As physicists close in on the Higgs boson, they should resist calls to change its name". *Nature*. 483, 374 (7390): 374. Bibcode:2012Natur.483..374.. doi:10.1038/483374a. Retrieved 21 January 2013.

[161] Becker, Kate (2012-03-29). "A Higgs by Any Other Name". "NOVA" (PBS) physics. Retrieved 21 January 2013.

[162] "Frequently Asked Questions: The Higgs!". *The Bulletin*. CERN. Retrieved 18 July 2012.

[163] Woit's physics blog *"Not Even Wrong"*: Anderson on Anderson-Higgs 2013-04-13

[164] Sample, Ian (2012-07-04). "Higgs boson's many great minds cause a Nobel prize headache". *The Guardian* (London). Retrieved 23 July 2013.

[165] "Rochester's Hagen Sakurai Prize Announcement" (Press release). University of Rochester. 2010.

[166] *C.R. Hagen Sakurai Prize Talk* (YouTube). 2010.

[167] Cho, A (2012-09-14). "Particle physics. Why the 'Higgs'?" (PDF). *Science* **337** (6100): 1287. doi:10.1126/science.337.6100.. PMID 22984044. Lee ... apparently used the term 'Higgs Boson' as early as 1966... but what may have made the term stick is a seminal paper Steven Weinberg...published in1967...Weinberg acknowledged the mix-up in an essay in the*New York Review of Books*in May2012. (See also the original article in*New York Review of Books*   [168] and Frank Close's 2011 book *The Infinity Puzzle*[79]:372)

[168] Weinberg, Steven (2012-05-10). "The Crisis of Big Science". *The New York Review of Books* (footnote 1). Retrieved 12 February 2013.

[169] Examples of early papers using the term "Higgs boson" include 'A phenomenological profile of the Higgs boson' (Ellis, Gaillard and Nanopoulos, 1976), 'Weak interaction theory and neutral currents' (Bjorken, 1977), and 'Mass of the Higgs boson' (Wienberg, received 1975)

[170] Leon Lederman; Dick Teresi (2006). *The God Particle: If the Universe Is the Answer, What Is the Question?*. Houghton Mifflin Harcourt. ISBN 0-547-52462-5.

[171] Kelly Dickerson (September 8, 2014). "Stephen Hawking Says 'God Particle' Could Wipe Out the Universe". livescience.com.

[172] Jim Baggott (2012). *Higgs: The invention and discovery of the 'God Particle'.* Oxford University Press. ISBN 978-0-19-165003-1.

[173] Scientific American Editors (2012). *The Higgs Boson: Searching for the God Particle.* Macmillan. ISBN 978-1-4668-2413-3.

[174] Ted Jaeckel (2007). *The God Particle: The Discovery and Modeling of the Ultimate Prime Particle.* Universal-Publishers. ISBN 978-1-58112-959-5.

[175] Aschenbach, Joy (1993-12-05). "No Resurrection in Sight for Moribund Super Collider : Science: Global financial partnerships could be the only way to salvage such a project. Some feel that Congress delivered a fatal blow". *Los Angeles Times*. Retrieved 16 January 2013. 'We have to keep the momentum and optimism and start thinking about international collaboration,' said Leon M. Lederman, the Nobel Prize-winning physicist who was the architect of the super collider plan

[176] "A Supercompetition For Illinois". *Chicago Tribune*. 1986-10-31. Retrieved 16 January 2013. The SSC, proposed by the U.S. Department of Energy in 1983, is a mind-bending project ... this gigantic laboratory ... this titanic project

[177] Diaz, Jesus (2012-12-15). "This Is [The] World's Largest Super Collider That Never Was". *Gizmodo*. Retrieved 16 January 2013. ...this titanic complex...

[178] Abbott, Charles (June 1987). "Illinois Issues journal, June 1987". p. 18. Lederman, who considers himself an unofficial propagandist for the super collider, said the SSC could reverse the physics brain drain in which bright young physicists have left America to work in Europe and elsewhere.

[179] Kevles, Dan. "Good-bye to the SSC: On the Life and Death of the Superconducting Super Collider" (PDF). *California Institute of Technology: "Engineering & Science".* 58 no. 2 (Winter 1995): 16–25. Retrieved 16 January 2013. Lederman, one of the principal spokesmen for the SSC, was an accomplished high-energy experimentalist who had made Nobel Prize-winning contributions to the development of the Standard Model during the 1960s (although the prize itself did not come until 1988). He was a fixture at congressional hearings on the collider, an unbridled advocate of its merits.

[180] Calder, Nigel (2005). *Magic Universe:A Grand Tour of Modern Science.* pp. 369–370. ISBN 9780191622359. The possibility that the next big machine would create the Higgs became a carrot to dangle in front of funding agencies and politicians. A prominent American physicist, Leon lederman *[sic]*, advertised the Higgs as The God Particle in the title of a book published in 1993 ...Lederman was involved in a campaign to persuade the US government to continue funding the Superconducting Super Collider... the ink was not dry on Lederman's book before the US Congress decided to write off the billions of dollars already spent

[181] Lederman, Leon (1993). *The God Particle If the Universe Is the Answer, What Is the Question?* (PDF). Dell Publishing. p. Chapter 2, Page 2. ISBN 0-385-31211-3. Retrieved 30 July 2015.

[182] Alister McGrath, Higgs boson: the particle of faith, *The Daily Telegraph*, Published 15 December 2011. Retrieved 15 December 2011.

[183] Sample, Ian (3 March 2009). "Father of the God particle: Portrait of Peter Higgs unveiled". London: The Guardian. Retrieved 24 June 2009.

[184] Chivers, Tom (2011-12-13). "How the 'God particle' got its name". *The Telegraph* (London). Retrieved 2012-12-03.

[185] Key scientist sure "God particle" will be found soon Reuters news story. 7 April 2008.

[186] "Interview: the man behind the 'God particle'", New Scientist 13 September 2008, pp. 44–5 (original interview in the Guardian: Father of the 'God Particle', June 30, 2008)

[187] Sample, Ian (2010). *Massive: The Hunt for the God Particle.* pp. 148–149 and 278–279. ISBN 9781905264957.

[188] Cole, K. (2000-12-14). "One Thing Is Perfectly Clear: Nothingness Is Perfect". *Los Angeles Times.* p. 'Science File'. Retrieved 17 January 2013. Consider the early universe–a state of pure, perfect nothingness; a formless fog of undifferentiated stuff ... 'perfect symmetry' ... What shattered this primordial perfection? One likely culprit is the so-called Higgs field ... Physicist Leon Lederman compares the way the Higgs operates to the biblical story of Babel [whose citizens] all spoke the same language ... Like God, says Lederman, the Higgs differentiated the perfect sameness, confusing everyone (physicists included) ... [Nobel Prizewinner Richard] Feynman wondered why the universe we live in was so obviously askew ... Perhaps, he speculated, total perfection would have been unacceptable to God. And so, just as God shattered the perfection of Babel, 'God made the laws only nearly symmetrical'

[189] Lederman, p. 22 *et seq*:

> "Something we cannot yet detect and which, one might say, has been put there to test and confuse us ... The issue is whether physicists will be confounded by this puzzle or whether, in contrast to the unhappy Babylonians, we will continue to build the tower and, as Einstein put it, 'know the mind of God'."
>
> "And the Lord said, Behold the people are un-confounding my confounding. And the Lord sighed and said, Go to, let us go down, and there give them the God Particle so that they may see how beautiful is the universe I have made".

[190] Sample, Ian (12 June 2009). "Higgs competition: Crack open the bubbly, the God particle is dead". *The Guardian* (London). Retrieved 4 May 2010.

[191] Gordon, Fraser (5 July 2012). "Introducing the higgson". *physicsworld.com*. Retrieved 25 August 2012.

[192] Wolchover, Natalie (2012-07-03). "Higgs Boson Explained: How 'God Particle' Gives Things Mass". *Huffington Post.* Retrieved 21 January 2013.

[193] Oliver, Laura (2012-07-04). "Higgs boson: how would you explain it to a seven-year-old?". *The Guardian* (London). Retrieved 21 January 2013.

[194] Zimmer, Ben (2012-07-15). "Higgs boson metaphors as clear as molasses". *The Boston Globe.* Retrieved 21 January 2013.

[195] "The Higgs particle: an analogy for Physics classroom (section)". www.lhc-closer.es (a collaboration website of LHCb physicist Xabier Vidal and High School Teachers at CERN educator Ramon Manzano). Retrieved 2013-01-09.

[196] Flam, Faye (2012-07-12). "Finally – A Higgs Boson Story Anyone Can Understand". *The Philadelphia Inquirer (philly.com).* Retrieved 21 January 2013.

[197] Sample, Ian (2011-04-28). "How will we know when the Higgs particle has been detected?". *The Guardian* (London). Retrieved 21 January 2013.

[198] Miller, David. "A quasi-political Explanation of the Higgs Boson; for Mr Waldegrave, UK Science Minister 1993". Retrieved 10 July 2012.

[199] Kathryn Grim. "Ten things you may not know about the Higgs boson". Symmetry Magazine. Retrieved 10 July 2012.

[200] David Goldberg, Associate Professor of Physics, Drexel University (2010-10-17). "What's the Matter with the Higgs Boson?". io9.com "Ask a physicist". Retrieved 21 January 2013.

[201] The Nobel Prize in Physics 1979 – official Nobel Prize website.

[202] The Nobel Prize in Physics 1999 – official Nobel Prize website.

[203] – official Nobel Prize website.

[204] Daigle, Katy (10 July 2012). "India: Enough about Higgs, let's discuss the boson". *AP News.* Retrieved 10 July 2012.

[205] Bal, Hartosh Singh (19 September 2012). "The Bose in the Boson". New York Times. Retrieved 21 September 2012.

[206] Alikhan, Anvar (16 July 2012). "The Spark In A Crowded Field". *Outlook India.* Retrieved 10 July 2012.

[207] Peskin & Schroeder 1995, Chapter 20

7.11   Further reading

- Nambu, Yoichiro; Jona-Lasinio, Giovanni (1961). "Dynamical Model of Elementary Particles Based on an Analogy with Superconductivity". *Physical Review* **122**: 345–358. Bibcode:1961PhRv..122..345N.doi:10.1103.

- Klein, Abraham; Lee, Benjamin W. (1964). "Does Spontaneous Breakdown of Symmetry Imply Zero-Mass Particles?". *Physical Review Letters* **12** (10): 266. Bibcode:1964PhRvL..12..266K. doi:10.1103/PhysRevLett.12.266.

- Anderson, Philip W. (1963). "Plasmons, Gauge Invariance, and Mass". *Physical Review* **130**: 439. Bibcode:1963. doi:10.1103/PhysRev.130.439.

- Gilbert, Walter (1964). "Broken Symmetries and Massless Particles". *Physical Review Letters* **12** (25): 713. Bibcode:1964PhRvL..12..713G. doi:10.1103/PhysRevLett.12.713.

- Higgs, Peter (1964). "Broken Symmetries, Massless Particles and Gauge Fields". *Physics Letters* **12** (2): 132–133. Bibcode:1964PhL....12..132H. doi:10.1016/0031-9163(64)91136-9.

- Guralnik, Gerald S.; Hagen, C.R.; Kibble, Tom W.B. (1968). "Broken Symmetries and the Goldstone Theorem". In R.L. Cool and R.E. Marshak. *Advances in Physics, Vol. 2.* Interscience Publishers. pp. 567–708. ISBN 978-0470170571.

7.12   External links

7.12.1   Popular science, mass media, and general coverage

- Hunting the Higgs Boson at C.M.S. Experiment, at CERN

- The Higgs Boson" by the CERN exploratorium.

- "Particle Fever", documentary film about the search for the Higgs Boson.

- "The Atom Smashers", documentary film about the search for the Higgs Boson at Fermilab.

- Collected Articles at the *Guardian*

- Video (04:38) – CERN Announcement on 4 July 2012, of the discovery of a particle which is suspected will be a Higgs Boson.

- Video1 (07:44) + Video2 (07:44) – Higgs Boson Explained by CERN Physicist, Dr. Daniel Whiteson (16 June 2011).

- HowStuffWorks: What exactly is the Higgs Boson?

- Carroll, Sean. "Higgs Boson with Sean Carroll". *Sixty Symbols*. University of Nottingham.

- Overbye, Dennis (2013-03-05). "Chasing the Higgs Boson: How 2 teams of rivals at CERN searched for physics' most elusive particle". *New York Times Science pages*. Retrieved 22 July 2013. - New York Times "behind the scenes" style article on the Higgs' search at ATLAS and CMS

- The story of the Higgs theory by the authors of the PRL papers and others closely associated:

  - Higgs, Peter (2010). "My Life as a Boson" (PDF). Talk given at Kings College, London, Nov 24 2010. Retrieved 17 January 2013. (also:)

  - Kibble, Tom (2009). "Englert–Brout–Higgs–Guralnik–Hagen–Kibble mechanism (history)". Scholarpedia. Retrieved 17 January 2013. (also:)

- Guralnik, Gerald (2009). "The History of the Guralnik, Hagen and Kibble development of the Theory of Spontaneous Symmetry Breaking and Gauge Particles". *International Journal of Modern Physics A* **24** (14): 2601–2627. arXiv:0907.3466. Bibcode:2009IJMPA..24.2601G. doi:10.1142/S0217751X09045431., Guralnik, Gerald (2011). "The Beginnings of Spontaneous Symmetry Breaking in Particle Physics. Proceedings of the DPF-2011 Conference, Providence, RI, 8–13 August 2011". arXiv:1110.2253v1 [physics.hist-ph]., and Guralnik, Gerald (2013). "Heretical Ideas that Provided the Cornerstone for the Standard Model of Particle Physics". SPG MITTEILUNGEN March 2013, No. 39, (p. 14), and Talk at Brown University about the 1964 PRL papers

  - Philip Anderson (not one of the PRL authors) on symmetry breaking in superconductivity and its migration into particle physics and the PRL papers

- Cartoon about the search

- Cham, Jorge (2014-02-19). "True Tales from the Road: The Higgs Boson Re-Explained". *Piled Higher and Deeper*. Retrieved 2014-02-25.

7.12.2 Significant papers and other

- Observation of a new particle in the search for the Standard Model Higgs Boson with the ATLAS detector at the LHC

- Observation of a new Boson at a mass of 125 GeV with the CMS experiment at the LHC

- Particle Data Group: Review of searches for Higgs Bosons.

- 2001, a spacetime odyssey: proceedings of the Inaugural Conference of the Michigan Center for Theoretical Physics : Michigan, USA, 21–25 May 2001, (p.86 – 88), ed. Michael J. Duff, James T. Liu, ISBN 978-981-238-231-3, containing Higgs' story of the Higgs Boson.

- A.A. Migdal & A.M. Polyakov, *Spontaneous Breakdown of Strong Interaction Symmetry and the Absence of Massless Particles*, Sov.J.-JETP 24,91 (1966) - example of a 1966 Russian paper on the subject.

7.12.3 Introductions to the field

- Spontaneous symmetry breaking, gauge theories, the Higgs mechanism and all that (Bernstein, *Reviews of Modern Physics* Jan 1974) - an introduction of 47 pages covering the development, history and mathematics of Higgs theories from around 1950 to 1974.

Chapter 8

Graviton

This article is about the hypothetical particle. For other uses, see Graviton (disambiguation).

In physics, the **graviton** is a hypothetical elementary particle that mediates the force of gravitation in the framework of quantum field theory. If it exists, the graviton is expected to be massless (because the gravitational force appears to have unlimited range) and must be a spin-2 boson. The spin follows from the fact that the source of gravitation is the stress–energy tensor, a second-rank tensor (compared to electromagnetism's spin-1 photon, the source of which is the four-current, a first-rank tensor). Additionally, it can be shown that any massless spin-2 field would give rise to a force indistinguishable from gravitation, because a massless spin-2 field must couple to (interact with) the stress–energy tensor in the same way that the gravitational field does. Seeing as the graviton is hypothetical, its discovery would unite quantum theory with gravity.[4] This result suggests that, if a massless spin-2 particle is discovered, it must be the graviton, so that the only experimental verification needed for the graviton may simply be the discovery of a massless spin-2 particle.[5]

8.1 Theory

The four other known forces of nature are mediated by elementary particles: electromagnetism by the photon, the strong interaction by the gluons, the Higgs field by the Higgs Boson, and the weak interaction by the W and Z bosons. The hypothesis is that the gravitational interaction is likewise mediated by an – as yet undiscovered – elementary particle, dubbed as *the graviton*. In the classical limit, the theory would reduce to general relativity and conform to Newton's law of gravitation in the weak-field limit.[6][7][8]

8.1.1 Gravitons and renormalization

When describing graviton interactions, the classical theory (i.e., the tree diagrams) and semiclassical corrections (one-loop diagrams) behave normally, but Feynman diagrams with two (or more) loops lead to ultraviolet divergences; that is, infinite results that cannot be removed because the quantized general relativity is not renormalizable, unlike quantum electrodynamics. That is, the usual ways physicists calculate the probability that a particle will emit or absorb a graviton give nonsensical answers and the theory loses its predictive power. These problems, together with some conceptual puzzles, led many physicists to believe that a theory more complete than quantized general relativity must describe the behavior near the Planck scale.

8.1.2 Comparison with other forces

Unlike the force carriers of the other forces, gravitation plays a special role in general relativity in defining the spacetime in which events take place. In some descriptions, matter modifies the 'shape' of spacetime itself, and gravity is a result of this shape, an idea which at first glance may appear hard to match with the idea of a force acting between particles.[9]

Because the diffeomorphism invariance of the theory does not allow any particular space-time background to be singled out as the "true" space-time background, general relativity is said to be background independent. In contrast, the Standard Model is *not* background independent, with Minkowski space enjoying a special status as the fixed background space-time.[10] A theory of quantum gravity is needed in order to reconcile these differences.[11] Whether this theory should be background independent is an open question. The answer to this question will determine our understanding of what specific role gravitation plays in the fate of the universe.[12]

8.1.3   Gravitons in speculative theories

String theory predicts the existence of gravitons and their well-defined interactions. A graviton in perturbative string theory is a closed string in a very particular low-energy vibrational state. The scattering of gravitons in string theory can also be computed from the correlation functions in conformal field theory, as dictated by the AdS/CFT correspondence, or from matrix theory.

A feature of gravitons in string theory is that, as closed strings without endpoints, they would not be bound to branes and could move freely between them. If we live on a brane (as hypothesized by brane theories) this "leakage" of gravitons from the brane into higher-dimensional space could explain why gravitation is such a weak force, and gravitons from other branes adjacent to our own could provide a potential explanation for dark matter. However, if gravitons were to move completely freely between branes this would dilute gravity too much, causing a violation of Newton's inverse square law. To combat this, Lisa Randall found that a three-brane (such as ours) would have a gravitational pull of its own, preventing gravitons from drifting freely, possibly resulting in the diluted gravity we observe while roughly maintaining Newton's inverse square law.[13] See brane cosmology.

A theory by Ahmed Farag Ali and Saurya Das adds quantum mechanical corrections (using Bohm trajectories) to general relativistic geodesics. If gravitons are given a small but non-zero mass, it could explain the cosmological constant without need for dark energy and solve the smallness problem.[14]

8.2   Experimental observation

Unambiguous detection of individual gravitons, though not prohibited by any fundamental law, is impossible with any physically reasonable detector.[15] The reason is the extremely low cross section for the interaction of gravitons with matter. For example, a detector with the mass of Jupiter and 100% efficiency, placed in close orbit around a neutron star, would only be expected to observe one graviton every 10 years, even under the most favorable conditions. It would be impossible to discriminate these events from the background of neutrinos, since the dimensions of the required neutrino shield would ensure collapse into a black hole.[15]

However, experiments to detect gravitational waves, which may be viewed as coherent states of many gravitons, are underway (such as LIGO and VIRGO). Although these experiments cannot detect individual gravitons, they might provide information about certain properties of the graviton.[16] For example, if gravitational waves were observed to propagate slower than c (the speed of light in a vacuum), that would imply that the graviton has mass (however, gravitational waves must propagate slower than "c" in a region with non-zero mass density if they are to be detectable).[17] Astronomical observations of the kinematics of galaxies, especially the galaxy rotation problem and modified Newtonian dynamics, might point toward gravitons having non-zero mass.[18]

8.3   Difficulties and outstanding issues

Most theories containing gravitons suffer from severe problems. Attempts to extend the Standard Model or other quantum field theories by adding gravitons run into serious theoretical difficulties at high energies (processes involving energies close to or above the Planck scale) because of infinities arising due to quantum effects (in technical terms, gravitation is nonrenormalizable). Since classical general relativity and quantum mechanics seem to be incompatible at such energies, from a theoretical point of view, this situation is not tenable. One possible solution is to replace particles with strings.

String theories are quantum theories of gravity in the sense that they reduce to classical general relativity plus field theory at low energies, but are fully quantum mechanical, contain a graviton, and are believed to be mathematically consistent.[19]

8.4 See also

- Gravitomagnetism

- Gravitational wave

- Planck mass

- Gravitation

- Static forces and virtual-particle exchange

- Multiverse

- Gravitino

8.5 References

[1] G is used to avoid confusion with gluons (symbol g)

[2] Rovelli, C. (2001). "Notes for a brief history of quantum gravity". arXiv:gr-qc/0006061 [gr-qc].

[3] Blokhintsev, D. I.; Gal'perin, F. M. (1934). "Gipoteza neitrino i zakon sokhraneniya energii" [Neutrino hypothesis and conservation of energy]. *Pod Znamenem Marxisma* (in Russian) **6**: 147–157.

[4] Lightman, A. P.; Press, W. H.; Price, R. H.; Teukolsky, S. A. (1975). "Problem 12.16". *Problem book in Relativity and Gravitation*. Princeton University Press. ISBN 0-691-08162-X.

[5] For a comparison of the geometric derivation and the (non-geometric) spin-2 field derivation of general relativity, refer to box 18.1 (and also 17.2.5) of Misner, C. W.; Thorne, K. S.; Wheeler, J. A. (1973). *Gravitation*. W. H. Freeman. ISBN 0-7167-0344-0.

[6] Feynman, R. P.; Morinigo, F. B.; Wagner, W. G.; Hatfield, B. (1995). *Feynman Lectures on Gravitation*. Addison-Wesley. ISBN 0-201-62734-5.

[7] Zee, A. (2003). *Quantum Field Theory in a Nutshell*. Princeton University Press. ISBN 0-691-01019-6.

[8] Randall, L. (2005). *Warped Passages: Unraveling the Universe's Hidden Dimensions*. Ecco Press. ISBN 0-06-053108-8.

[9] See the other articles on General relativity, Gravitational field, Gravitational wave, etc

[10] Colosi, D.; et al. (2005). "Background independence in a nutshell: The dynamics of a tetrahedron". *Classical and Quantum Gravity* **22** (14): 2971. arXiv:gr-qc/0408079. Bibcode:2005CQGra..22.2971C. doi:10.1088/0264-9381/22/14/008.

[11] Witten, E. (1993). "Quantum Background Independence In String Theory". arXiv:hep-th/9306122 [hep-th].

[12] Smolin, L. (2005). "The case for background independence". arXiv:hep-th/0507235 [hep-th].

[13] Kaku, Michio (2006). *Parallel Worlds - The science of alternative universes and our future in the Cosmos*. pp. 218–221.

[14] Ali, Ahmed Farang (2014). "Cosmology from quantum potential". *Physical Letters B* **741**: 276–279. arXiv:1404.3093v3. doi:10.1016/j.physletb.2014.12.057.

[15] Rothman, T.; Boughn, S. (2006). "Can Gravitons be Detected?". *Foundations of Physics* **36** (12): 1801–1825. arXiv:gr-qc/0601043. Bibcode:2006FoPh...36.1801R. doi:10.1007/s10701-006-9081-9.

[16] Dyson, Freeman (8 October 2013). "Is a graviton detectable?". *International Journal of Modern Physics A* **28** (25): 1330041-1–1330035–14. Bibcode:2013IJMPA..2830041D. doi:10.1142/S0217751X1330041X.

[17] Will, C. M. (1998). "Bounding the mass of the graviton using gravitational-wave observations of inspiralling compact binaries". *Physical Review D* **57** (4): 2061–2068. arXiv:gr-qc/9709011. Bibcode:1998PhRvD..57.2061W. doi:10.1103/PhysRevD.57.2061.

[18] Trippe, S. (2013), "A Simplified Treatment of Gravitational Interaction on Galactic Scales", J. Kor. Astron. Soc. **46**, 41. arXiv:1211.4692

[19] Sokal, A. (July 22, 1996). "Don't Pull the String Yet on Superstring Theory". *The New York Times*. Retrieved March 26, 2010.

8.6 External links

-
- Graviton on *In Our Time* at the BBC. (listen now)

Chapter 9

Bose–Einstein condensate

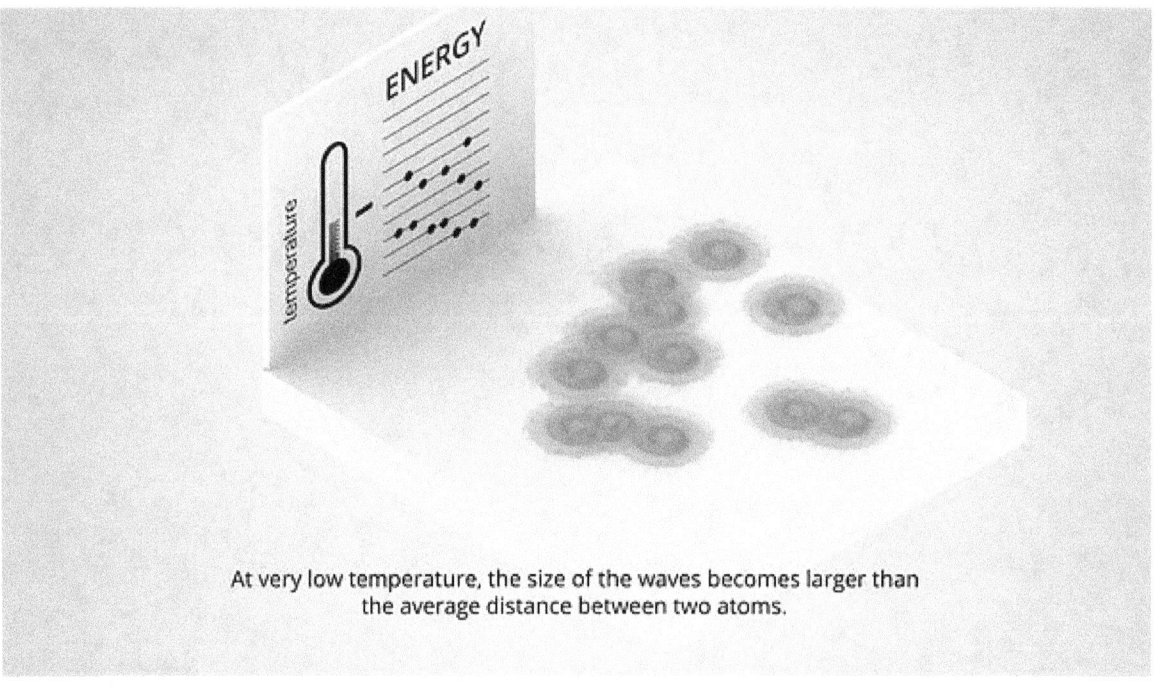

At very low temperature, the size of the waves becomes larger than the average distance between two atoms.

Schematic Bose-Einstein Condensation versus temperature and the energy diagram

A **Bose–Einstein condensate** (**BEC**) is a state of matter of a dilute gas of bosons cooled to temperatures very close to absolute zero (that is, very near 0 K or −273.15 °C). Under such conditions, a large fraction of bosons occupy the lowest quantum state, at which point macroscopic quantum phenomena become apparent.

This state was first predicted, generally, in 1924–25 by Satyendra Nath Bose and Albert Einstein.

9.1   History

Bose first sent a paper to Einstein on the quantum statistics of light quanta (now called photons). Einstein was impressed, translated the paper himself from English to German and submitted it for Bose to the *Zeitschrift für Physik*, which published it. (The Einstein manuscript, once believed to be lost, was found in a library at Leiden University in 2005.[1]). Einstein then extended Bose's ideas to matter in two other papers.[2] The result of their efforts is the concept of a Bose gas, governed by Bose–Einstein statistics, which describes the statistical distribution of identical particles with integer

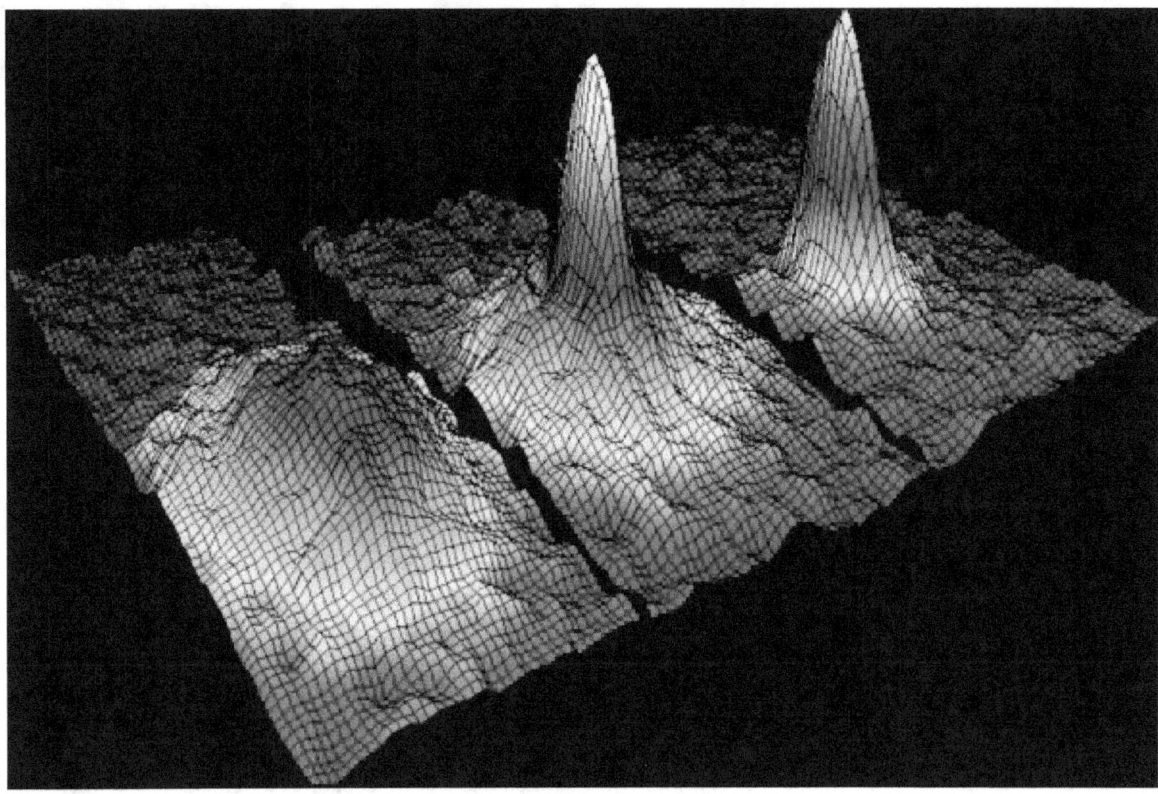

Velocity-distribution data (3 views) for a gas of rubidium atoms, confirming the discovery of a new phase of matter, the Bose–Einstein condensate. Left: just before the appearance of a Bose–Einstein condensate. Center: just after the appearance of the condensate. Right: after further evaporation, leaving a sample of nearly pure condensate.

spin, now called bosons. Bosons, which include the photon as well as atoms such as helium-4 (^{4}He), are allowed to share a quantum state. Einstein proposed that cooling bosonic atoms to a very low temperature would cause them to fall (or "condense") into the lowest accessible quantum state, resulting in a new form of matter.

In 1938, Fritz London proposed BEC as a mechanism for superfluidity in ^{4}He and superconductivity.[3][4]

On June 5, 1995, the first gaseous condensate was produced by Eric Cornell and Carl Wieman at the University of Colorado at Boulder NIST–JILA lab, in a gas of rubidium atoms cooled to 170 nanokelvin (nK).[5] Shortly thereafter, Wolfgang Ketterle at MIT demonstrated important BEC properties. For their achievements Cornell, Wieman, and Ketterle received the 2001 Nobel Prize in Physics.[6]

Many isotopes were soon condensed, then molecules, quasi-particles, and photons in 2010.[7]

9.2 Critical temperature

This transition to BEC occurs below a critical temperature, which for a uniform three-dimensional gas consisting of non-interacting particles with no apparent internal degrees of freedom is given by:

$$T_c = \left(\frac{n}{\zeta(3/2)}\right)^{2/3} \frac{2\pi\hbar^2}{mk_B} \approx 3.3125 \, \frac{\hbar^2 n^{2/3}}{mk_B}$$

where:

Interactions shift the value and the corrections can be calculated by mean-field theory.

9.3 Models

9.3.1 Einstein's non-interacting gas

Consider a collection of N noninteracting particles, which can each be in one of two quantum states, $|0\rangle$ and $|1\rangle$. If the two states are equal in energy, each different configuration is equally likely.

If we can tell which particle is which, there are 2^N different configurations, since each particle can be in $|0\rangle$ or $|1\rangle$ independently. In almost all of the configurations, about half the particles are in $|0\rangle$ and the other half in $|1\rangle$. The balance is a statistical effect: the number of configurations is largest when the particles are divided equally.

If the particles are indistinguishable, however, there are only $N+1$ different configurations. If there are K particles in state $|1\rangle$, there are $N - K$ particles in state $|0\rangle$. Whether any particular particle is in state $|0\rangle$ or in state $|1\rangle$ cannot be determined, so each value of K determines a unique quantum state for the whole system.

Suppose now that the energy of state $|1\rangle$ is slightly greater than the energy of state $|0\rangle$ by an amount E. At temperature T, a particle will have a lesser probability to be in state $|1\rangle$ by $e^{-E/kT}$. In the distinguishable case, the particle distribution will be biased slightly towards state $|0\rangle$. But in the indistinguishable case, since there is no statistical pressure toward equal numbers, the most-likely outcome is that most of the particles will collapse into state $|0\rangle$.

In the distinguishable case, for large N, the fraction in state $|0\rangle$ can be computed. It is the same as flipping a coin with probability proportional to $p = \exp(-E/T)$ to land tails.

In the indistinguishable case, each value of K is a single state, which has its own separate Boltzmann probability. So the probability distribution is exponential:

$$P(K) = Ce^{-KE/T} = Cp^K.$$

For large N, the normalization constant C is $(1 - p)$. The expected total number of particles not in the lowest energy state, in the limit that $N \rightarrow \infty$, is equal to $\sum_{n>0} Cnp^n = p/(1-p)$. It does not grow when N is large; it just approaches a constant. This will be a negligible fraction of the total number of particles. So a collection of enough Bose particles in thermal equilibrium will mostly be in the ground state, with only a few in any excited state, no matter how small the energy difference.

Consider now a gas of particles, which can be in different momentum states labeled $|k\rangle$. If the number of particles is less than the number of thermally accessible states, for high temperatures and low densities, the particles will all be in different states. In this limit, the gas is classical. As the density increases or the temperature decreases, the number of accessible states per particle becomes smaller, and at some point, more particles will be forced into a single state than the maximum allowed for that state by statistical weighting. From this point on, any extra particle added will go into the ground state.

To calculate the transition temperature at any density, integrate, over all momentum states, the expression for maximum number of excited particles, $p/(1 - p)$:

$$N = V \int \frac{d^3k}{(2\pi)^3} \frac{p(k)}{1 - p(k)} = V \int \frac{d^3k}{(2\pi)^3} \frac{1}{e^{\frac{k^2}{2mT}} - 1}$$

$$p(k) = e^{\frac{-k^2}{2mT}}.$$

When the integral is evaluated with factors of kB and $\hbar$ restored by dimensional analysis, it gives the critical temperature formula of the preceding section. Therefore, this integral defines the critical temperature and particle number corresponding to the conditions of negligible chemical potential. In Bose–Einstein statistics distribution, μ is actually still nonzero for BECs; however, μ is less than the ground state energy. Except when specifically talking about the ground state, μ can be approximated for most energy or momentum states as $\mu \approx 0$.

9.3.2 Bogoliubov theory for weakly interacting gas

Bogoliubov considered perturbations on the limit of dilute gas,[9] finding a finite pressure at zero temperature and positive chemical potential. This leads to corrections for the ground state. The Bogoliubov state has pressure(T=0): $P = g/2n^2$.

The original interacting system can be converted to a system of non-interacting particles with a dispersion law.

9.3.3 Gross–Pitaevskii equation

Main article: Gross–Pitaevskii equation

In some simplest cases, the state of condensed particles can be described with a nonlinear Schrödinger equation, also known as Gross-Pitaevskii or Ginzburg-Landau equation. The validity of this approach is actually limited to the case of ultracold temperatures, which fits well for the most alkali atoms experiments.

This approach originates from the assumption that the state of the BEC can be described by the unique wavefunction of the condensate $\psi(\vec{r})$. For a system of this nature, $|\psi(\vec{r})|^2$ is interpreted as the particle density, so the total number of atoms is $N = \int d\vec{r}|\psi(\vec{r})|^2$

Provided essentially all atoms are in the condensate (that is, have condensed to the ground state), and treating the bosons using mean field theory, the energy (E) associated with the state $\psi(\vec{r})$ is:

$$E = \int d\vec{r}\left[\frac{\hbar^2}{2m}|\nabla\psi(\vec{r})|^2 + V(\vec{r})|\psi(\vec{r})|^2 + \frac{1}{2}U_0|\psi(\vec{r})|^4\right]$$

Minimizing this energy with respect to infinitesimal variations in $\psi(\vec{r})$, and holding the number of atoms constant, yields the Gross–Pitaevski equation (GPE) (also a non-linear Schrödinger equation):

$$i\hbar\frac{\partial\psi(\vec{r})}{\partial t} = \left(-\frac{\hbar^2\nabla^2}{2m} + V(\vec{r}) + U_0|\psi(\vec{r})|^2\right)\psi(\vec{r})$$

where:

In the case of zero external potential, the dispersion law of interacting Bose-Einstein-condensed particles is given by so-called Bogoliubov spectrum (for $T = 0$):

$$\omega_p = \sqrt{\frac{p^2}{2m}\left(\frac{p^2}{2m} + 2U_0n_0\right)}$$

The Gross-Pitaevskii equation (GPE) provides a relatively good description of the behavior of atomic BEC's. However, GPE does not take into account the temperature dependence of dynamical variables, and is therefore valid only for $T = 0$. It is not applicable, for example, for the condensates of excitons, magnons and photons, where the critical temperature is up to room one.

Weaknesses of Gross–Pitaevskii model

The Gross–Pitaevskii model of BEC is a physical approximation valid for certain classes of BECs. By construction, the GPE uses the following simplifications: it assumes that interactions between condensate particles are of the contact two-body type and also neglects anomalous contributions to self-energy.[10] These assumptions are suitable mostly for the dilute three-dimensional condensates. If one relaxes any of these assumptions, the equation for the condensate wavefunction acquires the terms containing higher-order powers of the wavefunction. Moreover, for some physical systems the amount of such terms turns out to be infinite, therefore, the equation becomes essentially non-polynomial. The examples where this could happen are the Bose–Fermi composite condensates,[11][12][13][14] effectively lower-dimensional condensates,[15] and dense condensates and superfluid clusters and droplets.[16]

9.3.4   Other

However, it is clear that in a general case the behaviour of Bose–Einstein condensate can be described by coupled evolution equations for condensate density, superfluid velocity and distribution function of elementary excitations. This problem was in 1977 by Peletminskii et al. in microscopical approach. The Peletminskii equations are valid for any finite temperatures below the critical point. Years after, in 1985, Kirkpatrick and Dorfman obtained similar equations using another microscopical approach. The Peletminskii equations also reproduce Khalatnikov hydrodynamical equations for superfluid as a limiting case.

9.3.5   Superfluidity of BEC and Landau criterion

The phenomena of superfluidity of a Bose gas and superconductivity of a strongly-correlated Fermi gas (a gas of Cooper pairs) are tightly connected to Bose-Einstein condensation. Under corresponding conditions, below the temperature of phase transition, these phenomena were observed in helium-4 and different classes of superconductors. In this sense, the superconductivity is often called the superfluidity of Fermi gas. In the simplest form, the origin of superfluidity can be seen from the weakly interacting bosons model.

9.4   Experimental observation

9.4.1   Superfluid He-4

In 1938, Pyotr Kapitsa, John Allen and Don Misener discovered that helium-4 became a new kind of fluid, now known as a superfluid, at temperatures less than 2.17 K (the lambda point). Superfluid helium has many unusual properties, including zero viscosity (the ability to flow without dissipating energy) and the existence of quantized vortices. It was quickly believed that the superfluidity was due to partial Bose–Einstein condensation of the liquid. In fact, many properties of superfluid helium also appear in gaseous condensates created by Cornell, Wieman and Ketterle (see below). Superfluid helium-4 is a liquid rather than a gas, which means that the interactions between the atoms are relatively strong; the original theory of Bose–Einstein condensation must be heavily modified in order to describe it. Bose–Einstein condensation remains, however, fundamental to the superfluid properties of helium-4. Note that helium-3, a fermion, also enters a superfluid phase at low temperature, which can be explained by the formation of bosonic Cooper pairs of two atoms (see also fermionic condensate).

9.4.2   Gaseous

The first "pure" Bose–Einstein condensate was created by Eric Cornell, Carl Wieman, and co-workers at JILA on 5 June 1995. They cooled a dilute vapor of approximately two thousand rubidium-87 atoms to below 170 nK using a combination of laser cooling (a technique that won its inventors Steven Chu, Claude Cohen-Tannoudji, and William D. Phillips the 1997 Nobel Prize in Physics) and magnetic evaporative cooling. About four months later, an independent effort led by Wolfgang Ketterle at MIT condensed sodium-23. Ketterle's condensate had a hundred times more atoms, allowing important results such as the observation of quantum mechanical interference between two different condensates. Cornell, Wieman and Ketterle won the 2001 Nobel Prize in Physics for their achievements.[17]

A group led by Randall Hulet at Rice University announced a condensate of lithium atoms only one month following the JILA work.[18] Lithium has attractive interactions, causing the condensate to be unstable and collapse for all but a few atoms. Hulet's team subsequently showed the condensate could be stabilized by confinement quantum pressure for up to about 1000 atoms. Various isotopes have since been condensed.

Velocity-distribution data graph

In the image accompanying this article, the velocity-distribution data indicates the formation of a Bose–Einstein condensate out of a gas of rubidium atoms. The false colors indicate the number of atoms at each velocity, with red being the fewest and white being the most. The areas appearing white and light blue are at the lowest velocities. The peak is not infinitely narrow because of the Heisenberg uncertainty principle: spatially confined atoms have a minimum width velocity distribution. This width is given by the curvature of the magnetic potential in the given direction. More tightly confined directions have bigger widths in the ballistic velocity distribution. This anisotropy of the peak on the right is a purely quantum-mechanical effect and does not exist in the thermal distribution on the left. This graph served as the cover design for the 1999 textbook *Thermal Physics* by Ralph Baierlein.[19]

9.4.3 Quasiparticles

Main article: Bose-Einstein condensation of quasiparticles

Bose–Einstein condensation also applies to quasiparticles in solids. Magnons, Excitons, and Polaritons have integer spin and form condensates.

Magnons, electron spin waves, can be controlled by a magnetic field. Densities from the limit of a dilute gas to a strongly interacting Bose liquid are possible. Magnetic ordering is the analog of superfluidity. In 1999 condensation was demonstrated in antiferromagnetic $TlCuCl_3$,[20] at temperatures as large as 14 K. The high transition temperature (relative to atomic gases) is due to the magnons small mass (near an electron) and greater achievable density. In 2006, condensation in a ferromagnetic Yttrium-iron-garnet thin film was seen even at room temperature,[21][22] with optical pumping.

Excitons, electron-hole pairs, were predicted to condense at low temperature and high density by Boer et al. in 1961. Bilayer system experiments first demonstrated condensation in 2003, by Hall voltage disappearance. Fast optical exciton creation was used to form condensates in sub-Kelvin Cu_2O in 2005 on.

Polariton condensation was detected in a 5 K quantum well microcavity.

9.5 Peculiar properties

9.5.1 Vortices

As in many other systems, vortices can exist in BECs. These can be created, for example, by 'stirring' the condensate with lasers, or rotating the confining trap. The vortex created will be a quantum vortex. These phenomena are allowed for by the non-linear $|\psi(\vec{r})|^2$ term in the GPE. As the vortices must have quantized angular momentum the wavefunction may have the form $\psi(\vec{r}) = \phi(\rho, z)e^{i\ell\theta}$ where ρ, z and θ are as in the cylindrical coordinate system, and ℓ is the angular number. This is particularly likely for an axially symmetric (for instance, harmonic) confining potential, which is commonly used. The notion is easily generalized. To determine $\phi(\rho, z)$, the energy of $\psi(\vec{r})$ must be minimized, according to the constraint $\psi(\vec{r}) = \phi(\rho, z)e^{i\ell\theta}$. This is usually done computationally, however in a uniform medium the analytic form

$$\phi = \frac{nx}{\sqrt{2 + x^2}}$$

demonstrates the correct behavior, and is a good approximation.

A singly charged vortex ($\ell = 1$) is in the ground state, with its energy ϵ_v given by

$$\epsilon_v = \pi n \frac{\hbar^2}{m} \ln \left(1.464 \frac{b}{\xi} \right)$$

where b is the farthest distance from the vortex considered.(To obtain an energy which is well defined it is necessary to include this boundary b .)

For multiply charged vortices ($\ell > 1$) the energy is approximated by

$$\epsilon_v \approx \ell^2 \pi n \frac{\hbar^2}{m} \ln \left(\frac{b}{\xi} \right)$$

which is greater than that of ℓ singly charged vortices, indicating that these multiply charged vortices are unstable to decay. Research has, however, indicated they are metastable states, so may have relatively long lifetimes.

Closely related to the creation of vortices in BECs is the generation of so-called dark solitons in one-dimensional BECs. These topological objects feature a phase gradient across their nodal plane, which stabilizes their shape even in propagation and interaction. Although solitons carry no charge and are thus prone to decay, relatively long-lived dark solitons have been produced and studied extensively.[23]

9.5.2   Attractive interactions

Experiments led by Randall Hulet at Rice University from 1995 through 2000 showed that lithium condensates with attractive interactions could stably exist up to a critical atom number. Quench cooling the gas, they observed the condensate to grow, then subsequently collapse as the attraction overwhelmed the zero-point energy of the confining potential, in a burst reminiscent of a supernova, with an explosion preceded by an implosion.

Further work on attractive condensates was performed in 2000 by the JILA team, of Cornell, Wieman and coworkers. Their instrumentation now had better control so they used naturally *attracting* atoms of rubidium-85 (having negative atom–atom scattering length). Through Feshbach resonance involving a sweep of the magnetic field causing spin flip collisions, they lowered the characteristic, discrete energies at which rubidium bonds, making their Rb-85 atoms repulsive and creating a stable condensate. The reversible flip from attraction to repulsion stems from quantum interference among wave-like condensate atoms.

When the JILA team raised the magnetic field strength further, the condensate suddenly reverted to attraction, imploded and shrank beyond detection, then exploded, expelling about two-thirds of its 10,000 atoms. About half of the atoms in the condensate seemed to have disappeared from the experiment altogether, not seen in the cold remnant or expanding gas cloud.[17] Carl Wieman explained that under current atomic theory this characteristic of Bose–Einstein condensate could not be explained because the energy state of an atom near absolute zero should not be enough to cause an implosion; however, subsequent mean field theories have been proposed to explain it. Most likely they formed molecules of two rubidium atoms,[24] energy gained by this bond imparts velocity sufficient to leave the trap without being detected.

9.6   Current research

Compared to more commonly encountered states of matter, Bose–Einstein condensates are extremely fragile. The slightest interaction with the external environment can be enough to warm them past the condensation threshold, eliminating their interesting properties and forming a normal gas.

Nevertheless, they have proven useful in exploring a wide range of questions in fundamental physics, and the years since the initial discoveries by the JILA and MIT groups have seen an increase in experimental and theoretical activity. Examples include experiments that have demonstrated interference between condensates due to wave–particle duality,[25] the study of superfluidity and quantized vortices, the creation of bright matter wave solitons from Bose condensates confined to one dimension, and the slowing of light pulses to very low speeds using electromagnetically induced transparency.[26] Vortices in Bose–Einstein condensates are also currently the subject of analogue gravity research, studying the possibility of modeling black holes and their related phenomena in such environments in the laboratory. Experimenters have also realized "optical lattices", where the interference pattern from overlapping lasers provides a periodic potential. These have been used to explore the transition between a superfluid and a Mott insulator,[27] and may be useful in studying Bose–Einstein condensation in fewer than three dimensions, for example the Tonks–Girardeau gas.

Bose–Einstein condensates composed of a wide range of isotopes have been produced.[28]

Cooling fermions to extremely low temperatures has created degenerate gases, subject to the Pauli exclusion principle. To exhibit Bose–Einstein condensation, the fermions must "pair up" to form bosonic compound particles (e.g. molecules or Cooper pairs). The first molecular condensates were created in November 2003 by the groups of Rudolf Grimm at the University of Innsbruck, Deborah S. Jin at the University of Colorado at Boulder and Wolfgang Ketterle at MIT. Jin quickly went on to create the first fermionic condensate composed of Cooper pairs.[29]

In 1999, Danish physicist Lene Hau led a team from Harvard University which slowed a beam of light to about 17 meters per second., using a superfluid.[30] Hau and her associates have since made a group of condensate atoms recoil from a light pulse such that they recorded the light's phase and amplitude, recovered by a second nearby condensate, in what they term "slow-light-mediated atomic matter-wave amplification" using Bose–Einstein condensates: details are discussed in *Nature*.[31]

Researchers in the new field of atomtronics use the properties of Bose–Einstein condensates when manipulating groups of identical cold atoms using lasers.[32] Further, BECs have been proposed by Emmanuel David Tannenbaum for anti-stealth technology.[33]

9.6.1 Isotopes

The effect has mainly been observed on alkaline atoms which have nuclear properties particularly suitable for working with traps. As of 2012, using ultra-low temperatures of 10^{-7} K or below, Bose–Einstein condensates had been obtained for a multitude of isotopes, mainly of alkaline, alkaline earth, and lanthanoid atoms (^{7}Li, ^{23}Na, ^{39}K, ^{41}K, ^{85}Rb, ^{87}Rb, ^{133}Cs, ^{52}Cr, ^{40}Ca, ^{84}Sr, ^{86}Sr, ^{88}Sr, ^{174}Yb, ^{164}Dy, and ^{168}Er). Research was finally successful in hydrogen with aid of special methods. In contrast, the superfluid state of ^{4}He below 2.17 K is not a good example, because the interaction between the atoms is too strong. Only 8% of atoms are in the ground state near absolute zero, rather than the 100% of a true condensate.

The bosonic behavior of some of these alkaline gases appears odd at first sight, because their nuclei have half-integer total spin. It arises from a subtle interplay of electronic and nuclear spins: at ultra-low temperatures and corresponding excitation energies, the half-integer total spin of the electronic shell and half-integer total spin of the nucleus are coupled by a very weak hyperfine interaction. The total spin of the atom, arising from this coupling, is an integer lower value. The chemistry of systems at room temperature is determined by the electronic properties, which is essentially fermionic, since room temperature thermal excitations have typical energies much higher than the hyperfine values.

9.7   See also

- Atom laser

- Atomic coherence

- Bose–Einstein correlations

- Bose–Einstein condensation: a network theory approach

- Bose-Einstein condensation of excitons

- Cold Atom Laboratory

- Electromagnetically induced transparency

- Fermionic condensate

- Gas in a box

- Gross–Pitaevskii equation

- Macroscopic quantum phenomena

- Macroscopic quantum self-trapping

- Slow light

- Superconductivity

- Superfluid film

- Superfluid helium-4

- Supersolid

- Tachyon condensation

- Timeline of low-temperature technology

- Super-heavy atom

- Wiener sausage

9.8   References

[1] "Leiden University Einstein archive". Lorentz.leidenuniv.nl. 27 October 1920. Retrieved 23 March 2011.

[2] Clark, Ronald W. (1971). *Einstein: The Life and Times*. Avon Books. pp. 408–409. ISBN 0-380-01159-X.

[3] London, F. (1938). "The λ-Phenomenon of Liquid Helium and the Bose–Einstein Degeneracy". *Nature* **141** (3571): 643–644. Bibcode:1938Natur.141..643L. doi:10.1038/141643a0.

[4] London, F. *Superfluids* Vol.I and II, (reprinted New York: Dover 1964)

[5] http://www.nist.gov/public_affairs/releases/bec_background.cfm

[6] Levi, Barbara Goss (2001). "Cornell, Ketterle, and Wieman Share Nobel Prize for Bose–Einstein Condensates". *Search & Discovery*. Physics Today online. Archived from the original on 24 October 2007. Retrieved 26 January 2008.

[7] Klaers, Jan; Schmitt, Julian; Vewinger, Frank; Weitz, Martin (2010). "Bose–Einstein condensation of photons in an optical microcavity". *Nature* **468** (7323): 545–548. arXiv:1007.4088. Bibcode:2010Natur.468..545K. doi:10.1038/nature09567. PMID 21107426.

[8] (sequence A078434 in OEIS)

[9] N. N. Bogoliubov (1947). "On the theory of superfluidity.". *J. Phys. (USSR), 11:23*.

[10] Beliaev, S. T. Zh. Eksp. Teor. Fiz. 34, 418–432 (1958); ibid. 433–446 [Soviet Phys. JETP 3, 299 (1957)].

[11] Schick, M. (1971). "Two-Dimensional System of Hard-Core Bosons". *Physical Review A* **3** (3): 1067. Bibcode:1971PhRvA...3.1067S. doi:10.1103/PhysRevA.3.1067.

[12] Kolomeisky, E.; Straley, J. (1992). "Renormalization-group analysis of the ground-state properties of dilute Bose systems in d spatial dimensions". *Physical Review B* **46** (18): 11749. Bibcode:1992PhRvB..4611749K. doi:10.1103/PhysRevB.46.11749.

[13] Kolomeisky, E. B.; Newman, T. J.; Straley, J. P.; Qi, X. (2000). "Low-Dimensional Bose Liquids: Beyond the Gross-Pitaevskii Approximation". *Physical Review Letters* **85** (6): 1146–1149. arXiv:cond-mat/0002282. Bibcode:2000PhRvL..85.1146K. doi:10.1103/PhysRevLett.85.1146. PMID 10991498.

[14] Chui,S.;Ryzhov,V. (2004). "Collapse transition in mixtures of bosons and fermions".*Physical Review A***69**(4). BibcodeC. doi:10.1103/PhysRevA.69.043607.

[15] Salasnich, L.; Parola, A.; Reatto, L. (2002). "Effective wave equations for the dynamics of cigar-shaped and disk-shaped Bose condensates".*Phys.Rev.A***65**(4): 043614. arXiv:cond-mat/0201395. Bibcode:2002PhRvA..65d3614S.doi:10.1103/PhysRev.

[16] Avdeenkov, A. V.; Zloshchastiev, K. G. (2011). "Quantum Bose liquids with logarithmic nonlinearity: Self-sustainability and emergence of spatial extent". *J. Phys. B: At. Mol. Opt. Phys.* **44** (19): 195303. arXiv:1108.0847. Bibcode:2011JPhB...44s5303A. doi:10.1088/0953-4075/44/19/195303.

[17] "Eric A. Cornell and Carl E. Wieman — Nobel Lecture" (PDF). nobelprize.org.

[18] Bradley, C. C.; Sackett, C. A.; Tollett, J. J.; Hulet, R. G. (1995). "Evidence of Bose-Einstein Condensation in an Atomic Gas with Attractive Interactions"(PDF).*Physical review letters***75**(9): 1687–1690. Bibcode:1995PhRvL..75.1687B.doi:10.11 87.PMID10060366.

[19] Baierlein, Ralph (1999). *Thermal Physics.* Cambridge University Press. ISBN 0-521-65838-1.

[20] Nikuni, T.; Oshikawa, M.; Oosawa, A.; Tanaka, H. (1999). "Bose–Einstein Condensation of Dilute Magnons in TlCuCl$_3$". *Physical Review Letters* **84** (25): 5868–71. arXiv:cond-mat/9908118. Bibcode:2000PhRvL..84.5868N. doi:10.1103/PhysRev 8.PMID10991075.

[21] Demokritov, S.O.; Demidov, VE; Dzyapko, O; Melkov, GA; Serga, AA; Hillebrands, B; Slavin, AN (2006). "Bose–Einstein condensation of quasi-equilibrium magnons at room temperature under pumping". *Nature* **443** (7110): 430–433. Bibcode:2 .doi:10.1038/nature05117. PMID17006509.

[22] *Magnon Bose Einstein Condensation* made simple. Website of the "Westfählische Wilhelms Universität Münster" Prof.Demokritov. Retrieved 25 June 2012.

[23] Becker, Christoph; Stellmer, Simon; Soltan-Panahi, Parvis; Dörscher, Sören; Baumert, Mathis; Richter, Eva-Maria; Kronjäger, Jochen; Bongs, Kai; Sengstock, Klaus (2008). "Oscillations and interactions of dark and dark–bright solitons in Bose–Einstein condensates". *Nature Physics* **4** (6): 496–501. arXiv:0804.0544. Bibcode:2008NatPh...4..496B. doi:10.1038/nphys962.

[24] van Putten,M.H.P.M. (2010). "Pair condensates produced in bosenovae".*Physics Letters A***374**(33): 3346. Bibcode:2010PhLA. doi:10.1016/j.physleta.2010.06.020.

[25] Gorlitz, Axel. "Interference of Condensates (BEC@MIT)". Cua.mit.edu. Retrieved 13 October 2009.

[26] Dutton, Zachary; Ginsberg, Naomi S.; Slowe, Christopher and Hau, Lene Vestergaard (2004). "The art of taming light: ultra-slow and stopped light" (PDF). *Europhysics News* **35** (2): 33. Bibcode:2004ENews..35...33D. doi:10.1051/epn:2004201.

[27] "From Superfluid to Insulator: Bose–Einstein Condensate Undergoes a Quantum Phase Transition". Qpt.physics.harvard.edu. Retrieved 13 October 2009.

[28] "Ten of the best for BEC". Physicsweb.org. 1 June 2005.

[29] "Fermionic condensate makes its debut". Physicsweb.org. 28 January 2004.

[30] Cromie, William J. (18 February 1999). "Physicists Slow Speed of Light". The Harvard University Gazette. Retrieved 26 January 2008.

[31] Ginsberg, N. S.; Garner, S. R.; Hau, L. V. (2007). "Coherent control of optical information with matter wave dynamics". *Nature* **445** (7128): 623–626. doi:10.1038/nature05493. PMID 17287804.

[32] Weiss,P. (12February2000). "Atomtronics may be the new electronics".*Science News Online***157**(7): 104. doi:10.2307/40. Retrieved 12 February 2011.

[33] Tannenbaum, Emmanuel David (1970). "Gravimetric Radar: Gravity-based detection of a point-mass moving in a static background". arXiv:1208.2377 [physics.ins-det].

9.9 Further reading

- Bose,S.N. (1924). "Plancks Gesetz und Lichtquantenhypothese".*Zeitschrift für Physik***26**: 178. Bibcode:192B. doi:10.1007/BF01327326.

- Einstein, A. (1925). "Quantentheorie des einatomigen idealen Gases". *Sitzungsberichte der Preussischen Akademie der Wissenschaften* **1**: 3.,

- Landau, L. D. (1941). "The theory of Superfluity of Helium 111". *J. Phys. USSR* **5**: 71–90.

- L.Landau(1941). "Theory of the Superfluidity of Helium II".*Physical Review***60**(4): 356–358. Bibcode:1946L. doi:10.1103/PhysRev.60.356.

- M.H. Anderson, J.R. Ensher, M.R. Matthews, C.E. Wieman, and E.A. Cornell (1995). "Observation of Bose–Einstein Condensation in a Dilute Atomic Vapor". *Science* **269** (5221): 198–201. Bibcode:1995Sci...269..198A. doi:10.1126/science.269.5221.198. JSTOR 2888436. PMID 17789847.

- C. Barcelo, S. Liberati and M. Visser (2001). "Analogue gravity from Bose–Einstein condensates". *Classical and Quantum Gravity* **18** (6): 1137–1156. arXiv:gr-qc/0011026. Bibcode:2001CQGra..18.1137B. doi:10.1088/0264-9381/18/6/312.

- P.G. Kevrekidis, R. Carretero-Gonzlaez, D.J. Frantzeskakis and I.G. Kevrekidis (2006). "Vortices in Bose–Einstein Condensates: Some Recent Developments". *Modern Physics Letters B* **5** (33).

- K.B. Davis, M.-O. Mewes, M.R. Andrews, N.J. van Druten, D.S. Durfee, D.M. Kurn, and W. Ketterle (1995). "Bose–Einstein condensation in a gas of sodium atoms". *Physical Review Letters* **75** (22): 3969–3973. 9D.doi:10.1103/PhysRevLett.75.3969. PMID 10059782..

- D. S. Jin, J. R. Ensher, M. R. Matthews, C. E. Wieman, and E. A. Cornell (1996). "Collective Excitations of a Bose–Einstein Condensate in a Dilute Gas". *Physical Review Letters* **77** (3): 420–423. Bibcode:1996PhRvL..77..420J. doi:10.1103/PhysRevLett.77.420. PMID 10062808.

- M. R. Andrews, C. G. Townsend, H.-J. Miesner, D. S. Durfee, D. M. Kurn, and W. Ketterle (1997). "Observation of interference between two Bose condensates". *Science* **275** (5300): 637–641. doi:10.1126/science.275.5300.637. PMID 9005843..

- Eric A. Cornell and Carl E. Wieman (1998). "The Bose–Einstein Condensate". *Scientific American* **278** (3): 40–45. doi:10.1038/scientificamerican0398-40.

- M. R. Matthews, B. P. Anderson, P. C. Haljan, D. S. Hall, C. E. Wieman, and E. A. Cornell (1999). "Vortices in a Bose–Einstein Condensate". *Physical Review Letters* **83** (13): 2498–2501. arXiv:cond-mat/9908209. Bibcode:1999PhRvL..83.2498M. doi:10.1103/PhysRevLett.83.2498.

- E.A. Donley, N.R. Claussen, S.L. Cornish, J.L. Roberts, E.A. Cornell, and C.E. Wieman (2001). "Dynamics of collapsing and exploding Bose–Einstein condensates". *Nature* **412** (6844): 295–299. arXiv:cond-mat/0105019. Bibcode:2001Natur.412..295D. doi:10.1038/35085500. PMID 11460153.

- A. G. Truscott, K. E. Strecker, W. I. McAlexander, G. B. Partridge, and R. G. Hulet (2001). "Observation of Fermi Pressure in a Gas of Trapped Atoms". *Science* **291** (5513): 2570–2572. Bibcode:2001Sci...291.2570T. doi:10.1126/science.1059318. PMID 11283362.

- M. Greiner, O. Mandel, T. Esslinger, T. W. Hänsch, I. Bloch (2002). "Quantum phase transition from a super-fluid to a Mott insulator in a gas of ultracold atoms". *Nature* **415** (6867): 39–44. Bibcode:2002Natur.415...39G. doi:10.1038/415039a. PMID 11780110..

- S. Jochim, M. Bartenstein, A. Altmeyer, G. Hendl, S. Riedl, C. Chin, J. Hecker Denschlag, and R. Grimm (2003). "Bose–Einstein Condensation of Molecules". *Science* **302** (5653): 2101–2103. Bibcode:2003Sci...302.2101J. doi:10.1126/science.1093280. PMID 14615548.

- Markus Greiner, Cindy A. Regal and Deborah S. Jin (2003). "Emergence of a molecular Bose–Einstein condensate from a Fermi gas". *Nature* **426** (6966): 537–540. Bibcode:2003Natur.426..537G. doi:10.1038/nature02199. PMID 14647340.

- M. W. Zwierlein, C. A. Stan, C. H. Schunck, S. M. F. Raupach, S. Gupta, Z. Hadzibabic, and W. Ketterle (2003). "Observation of Bose–Einstein Condensation of Molecules". *Physical Review Letters* **91** (25): 250401. arXiv:cond-mat/0311617. Bibcode:2003PhRvL..91y0401Z. doi:10.1103/PhysRevLett.91.250401. PMID 14754098.

- C. A. Regal, M. Greiner, and D. S. Jin (2004). "Observation of Resonance Condensation of Fermionic Atom Pairs". *Physical Review Letters* **92** (4): 040403. arXiv:cond-mat/0401554. Bibcode:2004PhRvL..92d0403R. doi:10.1103/PhysRevLett.92.040403. PMID 14995356.

- C. J. Pethick and H. Smith, *Bose–Einstein Condensation in Dilute Gases*, Cambridge University Press, Cambridge, 2001.

- Lev P. Pitaevskii and S. Stringari, *Bose–Einstein Condensation*, Clarendon Press, Oxford, 2003.

- Mackie M, Suominen KA, Javanainen J., "Mean-field theory of Feshbach-resonant interactions in 85Rb condensates." Phys Rev Lett. 2002 Oct 28;89(18):180403.

9.10 External links

- Bose–Einstein Condensation 2009 Conference Bose–Einstein Condensation 2009 – Frontiers in Quantum Gases

- BEC Homepage General introduction to Bose–Einstein condensation

- Nobel Prize in Physics 2001 – for the achievement of Bose–Einstein condensation in dilute gases of alkali atoms, and for early fundamental studies of the properties of the condensates

- Physics Today: Cornell, Ketterle, and Wieman Share Nobel Prize for Bose–Einstein Condensates

- Bose–Einstein Condensates at JILA

- Atomcool at Rice University

- Alkali Quantum Gases at MIT

- Atom Optics at UQ

- Einstein's manuscript on the Bose–Einstein condensate discovered at Leiden University

- Bose–Einstein condensate on arxiv.org

- Bosons – The Birds That Flock and Sing Together

- Easy BEC machine – information on constructing a Bose–Einstein condensate machine.

- Verging on absolute zero – Cosmos Online

- Lecture by W Ketterle at MIT in 2001

- Bose–Einstein Condensation at NIST – NIST resource on BEC

Chapter 10

Bosonic field

In quantum field theory, a **bosonic field** is a quantum field whose quanta are bosons; that is, they obey Bose–Einstein statistics. Bosonic fields obey canonical commutation relations, as distinct from the canonical anticommutation relations obeyed by fermionic fields.

Examples include scalar fields, describing spin-0 particles such as the Higgs boson, and gauge fields, describing spin-1 particles such as the photon.

10.1   Basic properties

Free (non-interacting) bosonic fields obey canonical commutation relations. Those relations also hold for interacting bosonic fields in the interaction picture, where the fields evolve in time as if free and the effects of the interaction are encoded in the evolution of the states. It is these commutation relations that imply Bose–Einstein statistics for the field quanta.

10.2   Examples

Examples of bosonic fields include scalar fields, gauge fields, and symmetric 2-tensor fields, which are characterized by their covariance under Lorentz transformations and have spins 0, 1 and 2, respectively. Physical examples, in the same order, are the Higgs field, the photon field, and the graviton field. Of the last two, only the photon field can be quantized using the conventional methods of canonical or path integral quantization. This has led to the theory of quantum electrodynamics, one of the most successful theories in physics. Quantization of gravity, on the other hand, is a long standing problem that has led to development of theories such as string theory and loop quantum gravity.

10.3   Spin and statistics

The spin-statistics theorem implies that quantization of local, relativistic field theories in 3+1 dimensions may lead either to bosonic or fermionic quantum fields, i.e., fields obeying commutation or anti-commutation relations, according to whether they have integer or half-integer spin, respectively. Thus bosonic fields are one of the two theoretically possible types of quantum field, namely those corresponding to particles with integer spin.

In a non-relativistic many-body theory, the spin and the statistical properties of the quanta are not directly related. In fact, the commutation or anti-commutation relations are assumed based on whether the theory one intends to study corresponds to particles obeying Bose–Einstein or Fermi–Dirac statistics. In this context the spin remains an internal quantum number that is only phenomenologically related to the statistical properties of the quanta. Examples of non-relativistic bosonic fields include those describing cold bosonic atoms, such as Helium-4.

Such non-relativistic fields are not as fundamental as their relativistic counterparts: they provide a convenient 're-packaging' of the many-body wave function describing the state of the system, whereas the relativistic fields described above are a necessary consequence of the consistent union of relativity and quantum mechanics.

10.4   See also

- Quantum triviality

10.5   References

- Edwards, D. (1981). "The Mathematical Foundations of Quantum Field Theory: Fermions, Gauge Fields, and Super-symmetry, Part I: Lattice Field Theories", *International J. of Theor. Phys.*, Vol. 20, No. 7.

- Hoffmann, S.E. et alia (2008) 'Hybrid Phase-Space Simulation Method for Interacting Bose Fields'. *Physical Review A* Vol. 78, Issue 1.

- Peskin, M and Schroeder, D. (1995). *An Introduction to Quantum Field Theory*, Westview Press.

- Srednicki, Mark (2007). *Quantum Field Theory*, Cambridge University Press, ISBN 978-0-521-86449-7.

- Weinberg, Steven (1995). *The Quantum Theory of Fields*, (3 volumes) Cambridge University Press.

Chapter 11

List of particles

This is a list of the different types of particles found or believed to exist in the whole of the universe. For individual lists of the different particles, see the list below.

11.1 Elementary particles

Main article: Elementary particle

Elementary particles are particles with no measurable internal structure; that is, they are not composed of other particles. They are the fundamental objects of quantum field theory. Many families and sub-families of elementary particles exist. Elementary particles are classified according to their spin. Fermions have half-integer spin while bosons have integer spin. All the particles of the Standard Model have been experimentally observed, recently including the Higgs boson.[1][2]

11.1.1 Fermions

Main article: Fermion

Fermions are one of the two fundamental classes of particles, the other being bosons. Fermion particles are described by Fermi–Dirac statistics and have quantum numbers described by the Pauli exclusion principle. They include the quarks and leptons, as well as any composite particles consisting of an odd number of these, such as all baryons and many atoms and nuclei.

Fermions have half-integer spin; for all known elementary fermions this is $1/2$. All known fermions, except neutrinos, are also Dirac fermions; that is, each known fermion has its own distinct antiparticle. It is not known whether the neutrino is a Dirac fermion or a Majorana fermion.[3] Fermions are the basic building blocks of all matter. They are classified according to whether they interact via the color force or not. In the Standard Model, there are 12 types of elementary fermions: six quarks and six leptons.

Quarks

Main article: Quark

Quarks are the fundamental constituents of hadrons and interact via the strong interaction. Quarks are the only known carriers of fractional charge, but because they combine in groups of three (baryons) or in groups of two with antiquarks (mesons), only integer charge is observed in nature. Their respective antiparticles are the antiquarks, which are identical except for the fact that they carry the opposite electric charge (for example the up quark carries charge $+^2/_3$, while the up antiquark carries charge $-^2/_3$), color charge, and baryon number. There are six flavors of quarks; the three positively charged quarks are called "up-type quarks" and the three negatively charged quarks are called "down-type quarks".

Leptons

Main article: Leptons

Leptons do not interact via the strong interaction. Their respective antiparticles are the antileptons which are identical, except for the fact that they carry the opposite electric charge and lepton number. The antiparticle of an electron is an antielectron, which is nearly always called a "positron" for historical reasons. There are six leptons in total; the three charged leptons are called "electron-like leptons", while the neutral leptons are called "neutrinos". Neutrinos are known to oscillate, so that neutrinos of definite flavor do not have definite mass, rather they exist in a superposition of mass eigenstates. The hypothetical heavy right-handed neutrino, called a "sterile neutrino", has been left off the list.

11.1.2 Bosons

Main article: Boson

Bosons are one of the two fundamental classes of particles, the other being fermions. Bosons are characterized by Bose–Einstein statistics and all have integer spins. Bosons may be either elementary, like photons and gluons, or composite, like mesons.

The fundamental forces of nature are mediated by gauge bosons, and mass is believed to be created by the Higgs field. According to the Standard Model the elementary bosons are:

The graviton is added to the list although it is not predicted by the Standard Model, but by other theories in the framework of quantum field theory. Furthermore, gravity is non-renormalizable. There are a total of eight independent gluons. The Higgs boson is postulated by the electroweak theory primarily to explain the origin of particle masses. In a process known as the "Higgs mechanism", the Higgs boson and the other gauge bosons in the Standard Model acquire mass via spontaneous symmetry breaking of the SU(2) gauge symmetry. The Minimal Supersymmetric Standard Model (MSSM) predicts several Higgs bosons. A new particle expected to be the Higgs boson was observed at the CERN/LHC on March 14, 2013, around the energy of 126.5GeV with an accuracy of close to five sigma (99.9999%, which is accepted as definitive). The Higgs mechanism giving mass to other particles has not been observed yet.

11.1.3 Hypothetical particles

Supersymmetric theories predict the existence of more particles, none of which have been confirmed experimentally as of 2014:

Note: just as the photon, Z boson and $W^\pm$ bosons are superpositions of the B^0, W^0, W^1, and W^2 fields – the photino, zino, and wino$^\pm$ are superpositions of the bino^0, wino^0, wino^1, and wino^2 by definition.

No matter if one uses the original gauginos or this superpositions as a basis, the only predicted physical particles are neutralinos and charginos as a superposition of them together with the Higgsinos.

Other theories predict the existence of additional bosons:

Mirror particles are predicted by theories that restore parity symmetry.

"Magnetic monopole" is a generic name for particles with non-zero magnetic charge. They are predicted by some GUTs.

"Tachyon" is a generic name for hypothetical particles that travel faster than the speed of light and have an imaginary rest mass.

Preons were suggested as subparticles of quarks and leptons, but modern collider experiments have all but ruled out their existence.

Kaluza–Klein towers of particles are predicted by some models of extra dimensions. The extra-dimensional momentum is manifested as extra mass in four-dimensional spacetime.

11.2   Composite particles

11.2.1   Hadrons

Main article: Hadron

Hadrons are defined as strongly interacting composite particles. Hadrons are either:

- Composite fermions, in which case they are called baryons.
- Composite bosons, in which case they are called mesons.

Quark models, first proposed in 1964 independently by Murray Gell-Mann and George Zweig (who called quarks "aces"), describe the known hadrons as composed of valence quarks and/or antiquarks, tightly bound by the color force, which is mediated by gluons. A "sea" of virtual quark-antiquark pairs is also present in each hadron.

Baryons

See also: List of baryons

Ordinary baryons (composite fermions) contain three valence quarks or three valence antiquarks each.

- Nucleons are the fermionic constituents of normal atomic nuclei:
    - Protons, composed of two up and one down quark (uud)
    - Neutrons, composed of two down and one up quark (ddu)
- Hyperons, such as the Λ, Σ, Ξ, and Ω particles, which contain one or more strange quarks, are short-lived and heavier than nucleons. Although not normally present in atomic nuclei, they can appear in short-lived hypernuclei.
- A number of charmed and bottom baryons have also been observed.

Some hints at the existence of exotic baryons have been found recently; however, negative results have also been reported. Their existence is uncertain.

- Pentaquarks consist of four valence quarks and one valence antiquark.

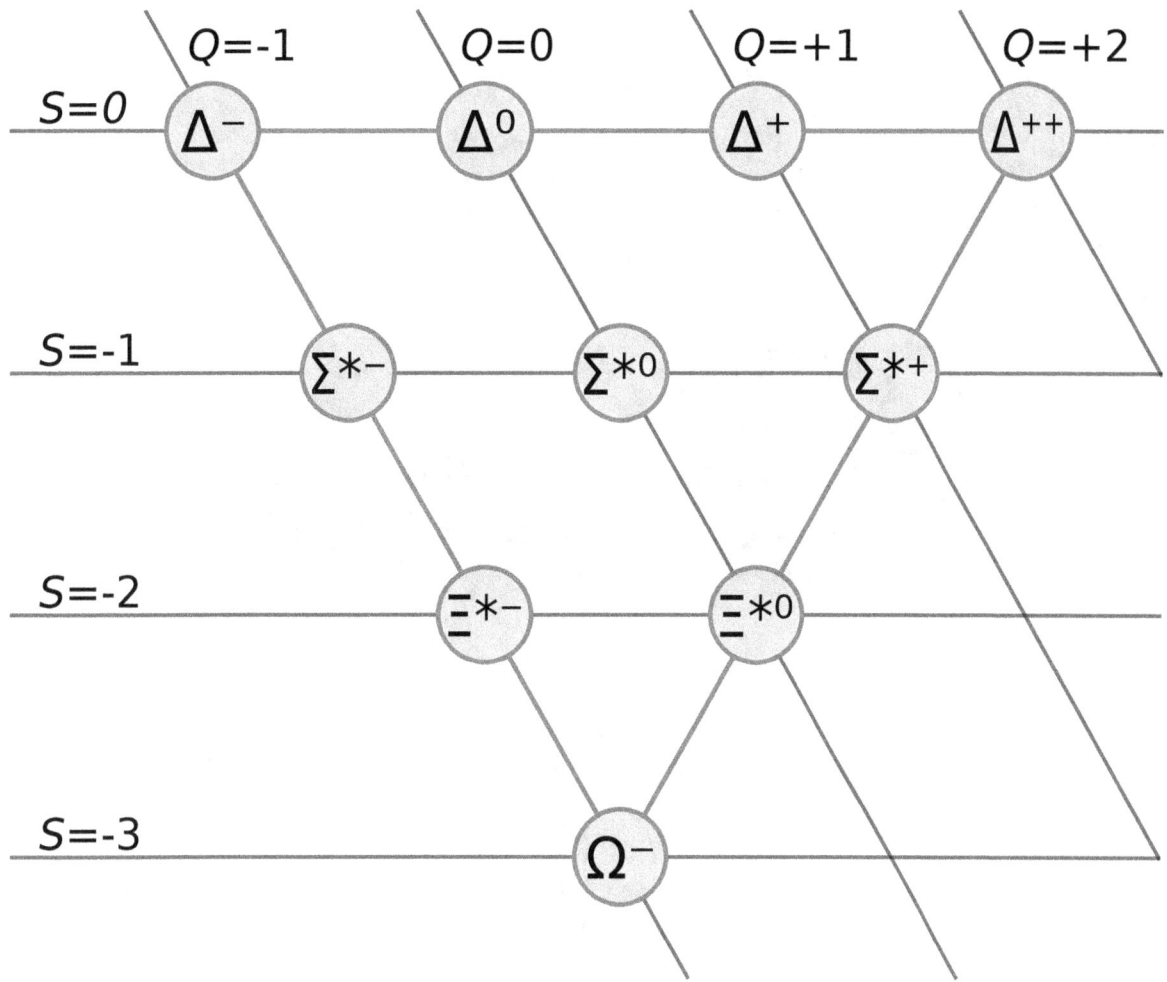

A combination of three u, d or s quarks with a total spin of ³⁄₂ form the so-called "baryon decuplet".

Mesons

See also: List of mesons

Ordinary mesons are made up of a valence quark and a valence antiquark. Because mesons have spin of 0 or 1 and are not themselves elementary particles, they are "composite" bosons. Examples of mesons include the pion, kaon, and the J/ψ. In quantum hydrodynamic models, mesons mediate the residual strong force between nucleons.

At one time or another, positive signatures have been reported for all of the following exotic mesons but their existences have yet to be confirmed.

- A tetraquark consists of two valence quarks and two valence antiquarks;

- A glueball is a bound state of gluons with no valence quarks;

- Hybrid mesons consist of one or more valence quark-antiquark pairs and one or more real gluons.

11.2.2 Atomic nuclei

Atomic nuclei consist of protons and neutrons. Each type of nucleus contains a specific number of protons and a specific

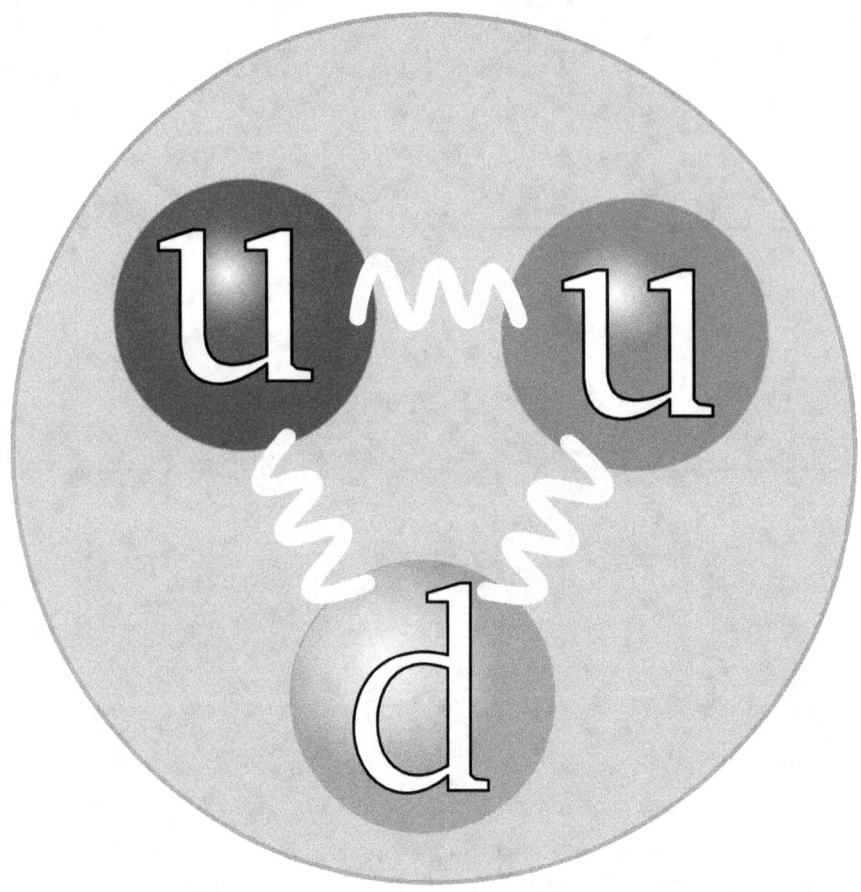

Proton quark structure: 2 up quarks and 1 down quark. The gluon tubes or flux tubes are now known to be Y shaped.

number of neutrons, and is called a "nuclide" or "isotope". Nuclear reactions can change one nuclide into another. See table of nuclides for a complete list of isotopes.

11.2.3 Atoms

Atoms are the smallest neutral particles into which matter can be divided by chemical reactions. An atom consists of a small, heavy nucleus surrounded by a relatively large, light cloud of electrons. Each type of atom corresponds to a specific chemical element. To date, 118 elements have been discovered, while only the elements 1-112,114, and 116 have received official names.

The atomic nucleus consists of protons and neutrons. Protons and neutrons are, in turn, made of quarks.

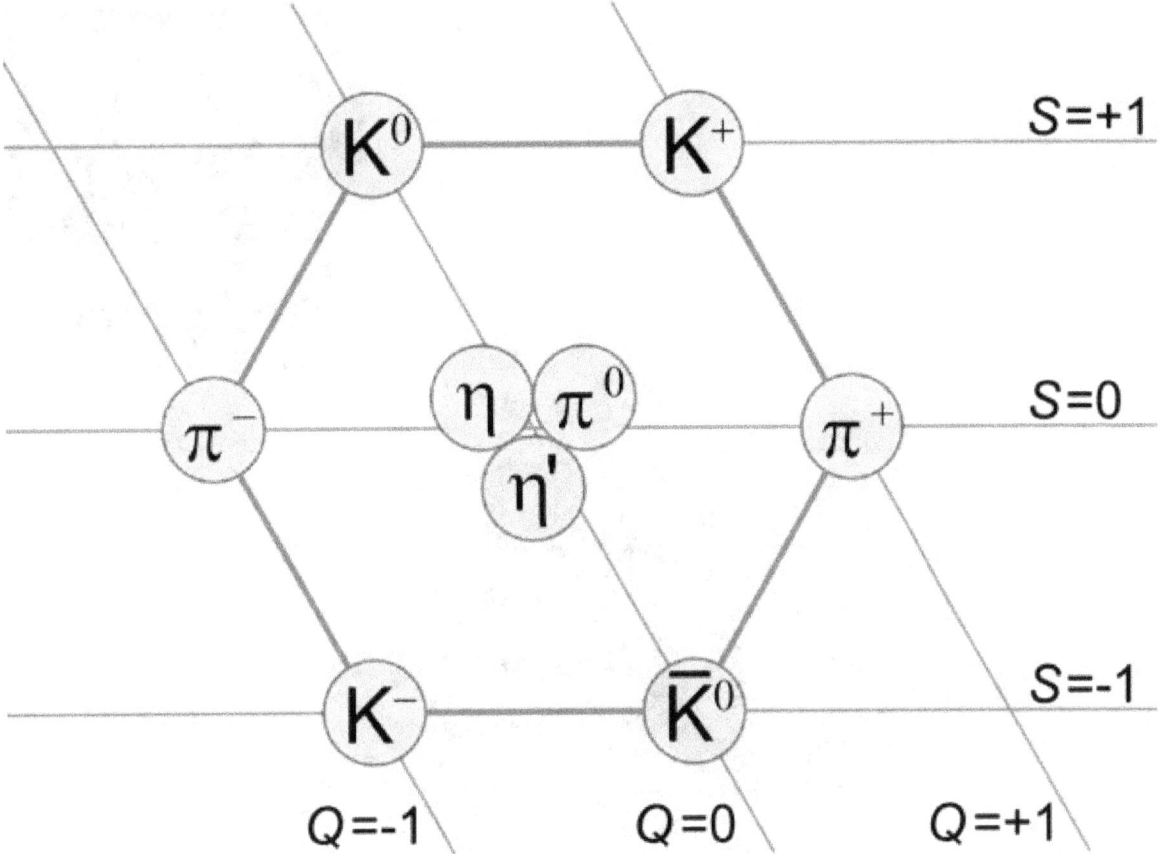

Mesons of spin 0 form a nonet

11.2.4 Molecules

Molecules are the smallest particles into which a non-elemental substance can be divided while maintaining the physical properties of the substance. Each type of molecule corresponds to a specific chemical compound. Molecules are a composite of two or more atoms. See list of compounds for a list of molecules.

11.3 Condensed matter

The field equations of condensed matter physics are remarkably similar to those of high energy particle physics. As a result, much of the theory of particle physics applies to condensed matter physics as well; in particular, there are a selection of field excitations, called quasi-particles, that can be created and explored. These include:

- Phonons are vibrational modes in a crystal lattice.

- Excitons are bound states of an electron and a hole.

- Plasmons are coherent excitations of a plasma.

- Polaritons are mixtures of photons with other quasi-particles.

- Polarons are moving, charged (quasi-) particles that are surrounded by ions in a material.

- Magnons are coherent excitations of electron spins in a material.

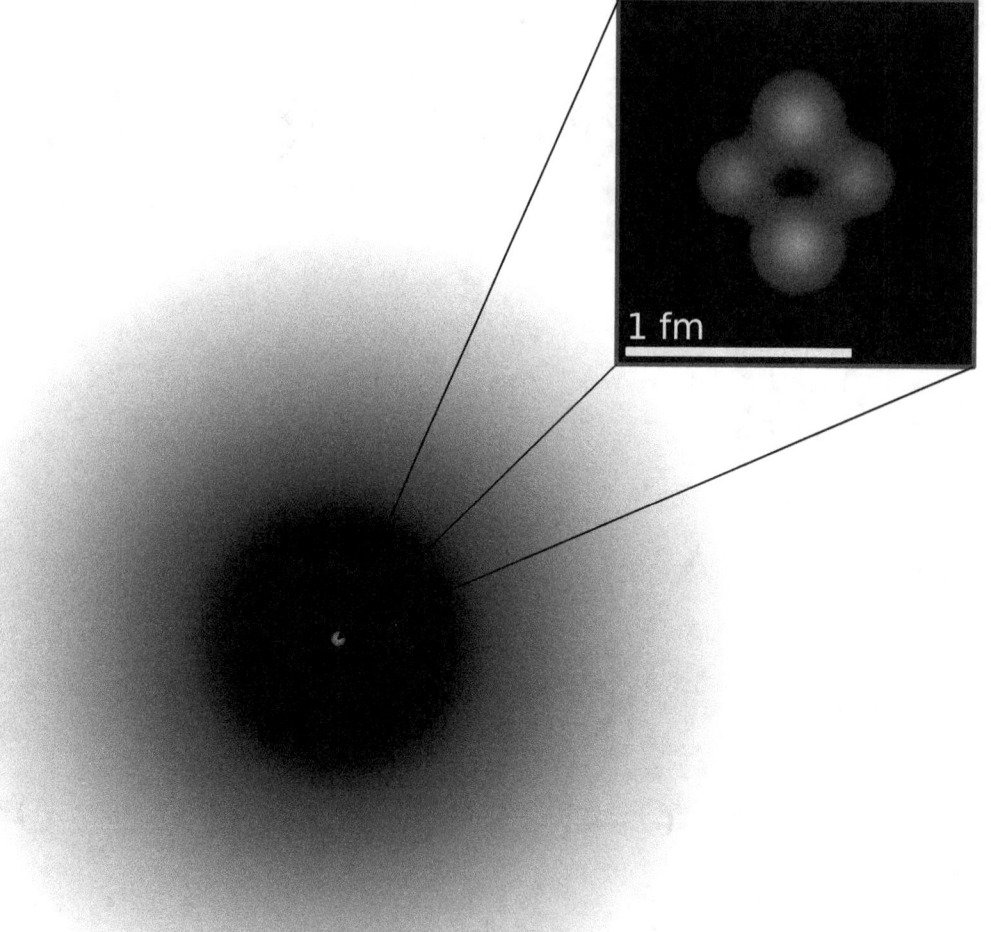

$$1 \text{ Å} = 100,000 \text{ fm}$$

A semi-accurate depiction of the helium atom. In the nucleus, the protons are in red and neutrons are in purple. In reality, the nucleus is also spherically symmetrical.

11.4   Other

- An anyon is a generalization of fermion and boson in two-dimensional systems like sheets of graphene that obeys braid statistics.

- A plekton is a theoretical kind of particle discussed as a generalization of the braid statistics of the anyon to dimension > 2.

- A WIMP (weakly interacting massive particle) is any one of a number of particles that might explain dark matter (such as the neutralino or the axion).

- The pomeron, used to explain the elastic scattering of hadrons and the location of Regge poles in Regge theory.

- The skyrmion, a topological solution of the pion field, used to model the low-energy properties of the nucleon, such as the axial vector current coupling and the mass.

- A genon is a particle existing in a closed timelike world line where spacetime is curled as in a Frank Tipler or Ronald Mallett time machine.

- A goldstone boson is a massless excitation of a field that has been spontaneously broken. The pions are quasi-goldstone bosons (quasi- because they are not exactly massless) of the broken chiral isospin symmetry of quantum chromodynamics.

- A goldstino is a goldstone fermion produced by the spontaneous breaking of supersymmetry.

- An instanton is a field configuration which is a local minimum of the Euclidean action. Instantons are used in nonperturbative calculations of tunneling rates.

- A dyon is a hypothetical particle with both electric and magnetic charges.

- A geon is an electromagnetic or gravitational wave which is held together in a confined region by the gravitational attraction of its own field energy.

- An inflaton is the generic name for an unidentified scalar particle responsible for the cosmic inflation.

- A spurion is the name given to a "particle" inserted mathematically into an isospin-violating decay in order to analyze it as though it conserved isospin.

- What is called "true muonium", a bound state of a muon and an antimuon, is a theoretical exotic atom which has never been observed.

11.5 Classification by speed

- A tardyon or bradyon travels slower than light and has a non-zero rest mass.

- A luxon travels at the speed of light and has no rest mass.

- A tachyon (mentioned above) is a hypothetical particle that travels faster than the speed of light and has an imaginary rest mass.

11.6 See also

- Acceleron

- List of baryons

- List of compounds for a list of molecules.

- List of fictional elements, materials, isotopes and atomic particles

- List of mesons

- Periodic table for an overview of atoms.

- Standard Model for the current theory of these particles.

- Table of nuclides

- Timeline of particle discoveries

11.7 References

[1] Observation of a new boson at a mass of 125 GeV with the CMS experiment at the LHC (2013). *arXiv:1207.7235.*

[2] Observation of a new particle in the search for the Standard Model Higgs boson with the ATLAS detector at the LHC (2012). *arXiv:1207.7214.*

[3] B. Kayser, *Two Questions About Neutrinos*, arXiv:1012.4469v1 [hep-ph] (2010).

[4] R. Maartens (2004). *Brane-World Gravity* (PDF). *Living Reviews in Relativity* **7**. p. 7. Also available in web format at http://www.livingreviews.org/lrr-2004-7.

- C. Amsler *et al.* (Particle Data Group) (2008). "Review of Particle Physics". *Physics Letters B* **667** (1–5): 1. Bibcode:2008PhLB..667....1P. doi:10.1016/j.physletb.2008.07.018. *(All information on this list, and more, can be found in the extensive, biannually-updated review by the Particle Data Group)*

11.8 Text and image sources, contributors, and licenses

11.8.1 Text

- **Boson** *Source:* https://en.wikipedia.org/wiki/Boson?oldid=682396358 *Contributors:* CYD, The Anome, Xaonon, Aldie, Enchanter, Roadrunner, Ben-Zin~enwiki, Lisiate, Michael Hardy, Tim Starling, Kroose, Looxix~enwiki, Andrewa, Glenn, Andres, Kaihsu, Samw, Panoramix, Schneelocke, CAkira, Wikiborg, The Anomebot, Saltine, Phys, Drxenocide, Robbot, Altenmann, Bkalafut, Merovingian, Rorro, Hadal, Robinh, VanishedUser kfljdfjsg33k, Giftlite, Fropuff, Pharotic, Isidore, Alexf, Beland, Jossi, Icairns, Zfr, Cructacean, Ornil, Mormegil, DanielCD, Noisy, Discospinster, Rich Farmbrough, Guanabot, Hidaspal, Sunborn, Kbh3rd, Jensbn, Alxndr, La goutte de pluie, Anthony Appleyard, Jlandahl, Leoadec, H2g2bob, Rocastelo, Benbest, Mpatel, Nakos2208~enwiki, GregorB, SDC, Palica, Ashmoo, Graham87, Kbdank71, Zzedar, Drbogdan, Strait, Master Justin, Wragge, FlaBot, Srleffler, Chobot, YurikBot, Bhny, The1physicist, Salsb, NawlinWiki, Welsh, Pyg, Dna-webmaster, Enormousdude, NeilN, Finell, Hal peridol, SmackBot, Incnis Mrsi, Ashley thomas80, Melchoir, Gilliam, MK8, MalafayaBot, Complexica, Epastore, DHN-bot~enwiki, Scienz Guy, Sbharris, QFT, Voyajer, Grover cleveland, Philvarner, Bradenripple, SashatoBot, Lambiam, Turbothy, T-dot, MagnaMopus, Candamir, WhiteHatLurker, Dicklyon, Treyp, Focomoso, Dan Gluck, UltraHighVacuum, Iridescent, Mathninja, Buckyboy314, Ianji, Cydebot, Stebbins, W.F.Galway, VashiDonsk, Tenbergen, Ward3001, Abtvctkto61, Thijs!bot, Barticus88, Mbell, Frozenport, Headbomb, MichaelMaggs, Escarbot, Orionus, Shan23, Alomas, JAnDbot, Deflective, CosineKitty, Pkoppenb, TheEditrix2, Magioladitis, VoABot II, Inertiatic076, Vanished user ty12kl89jq10, CodeCat, MartinBot, STBot, R'n'B, Tarotcards, Uberdude85, RuneSylvester, The Wild Falcon, Asnr 6, TXiKiBoT, Hqb, Anonymous Dissident, Abdullais4u, Bertrem, Moutane, Dirkbb, Antixt, Jeraaldo, BriEnBest, SieBot, Jim E. Black, Gerakibot, RadicalOne, Flyer22, Radon210, Sunayanaa, Jojalozzo, Tpvibes, Nsajjansajja, Owhanow~enwiki, Mike2vil, Mgurgan, VanishedUser sdu9aya9fs787sads, Danthewhale, PipepBot, Rodhullandemu, ChandlerMapBot, Excirial, PixelBot, Nilradical, Cenarium, Wikeepedian, Ouchitburns, Addbot, Bwr6, Minami Kana, Aboctok, Numbo3-bot, OlEnglish, David0811, WikiDreamer Bot, Jack who built the house, Luckas-bot, Yobot, Ptbotgourou, Senator Palpatine, AnomieBOT, Jim1138, JackieBot, Kingpin13, Materialscientist, Xqbot, Gravitivistically, Daners, Tomwsulcer, GrouchoBot, MeDrewNotYou, ⁇⁇, Ace of Spades, Alarics, Paul Laroque, Rameshngbot, RedBot, FoxBot, DixonDBot, Michael9422, Weedwhacker128, Tbhotch, TjBot, Ripchip Bot, EmausBot, JSquish, Kkm010, HiW-Bot, ZéroBot, StringTheory11, Lagomen, Robhenry9, Tls60, RockMagnetist, ClueBot NG, Raghavankl, GioGziro95, HBook, Helpful Pixie Bot, Bibcode Bot, 2001:db8, AvocatoBot, Nickni28, Minsbot, Blogger 20, Protomaestro, Abitoby, Darryl from Mars, NoRwEgIaNbAcTeRiUm, Jason7898, Valluvan888, Ov.kulkarni, Crpandya, Enamex, Lugia2453, Graphium, Federicoaolivieri, 77Mike77, Rltb, 314Username, Dllaughingwang, Codeusirae, Sometree, DR ROBERT HALT, KasparBot, Jiswin1992 and Anonymous: 223

- **Gauge boson** *Source:* https://en.wikipedia.org/wiki/Gauge_boson?oldid=682924361 *Contributors:* Bryan Derksen, Andre Engels, Michael Hardy, Ahoerstemeier, Bueller 007, LouI, Phys, Robbot, Gwrede, Rholton, Rursus, Davidl9999, Giftlite, Xerxes314, Alison, JeffBobFrank, Chinasaur, Andris, Garth 187, Beland, Setokaiba, Icairns, AmarChandra, Lumidek, Vsmith, Roybb95~enwiki, Mal~enwiki, La goutte de pluie, Nk, Kusma, Ringbang, Mpatel, Nakos2208~enwiki, Tevatron~enwiki, Kbdank71, Chobot, Roboto de Ajvol, Hairy Dude, Salsb, StuRat, ArielGold, RG2, InverseHypercube, Niels Olson, Sadi Carnot, TriTertButoxy, Ekjon Lok, Bjankuloski06en~enwiki, Phatom87, Headbomb, Tyco.skinner, Knotwork, Swpb, Maurice Carbonaro, Gombang, TXiKiBoT, Odellus, Antixt, AlleborgoBot, SieBot, Jim E. Black, Homonihilis, BOTarate, DumZiBoT, SilvonenBot, Addbot, Bertman600, NjardarBot, Numbo3-bot, Lightbot, Zorrobot, Luckas-bot, Yobot, Citation bot, ArthurBot, A. di M., Rameshngbot, RedBot, RobinK, Mary at CERN, TjBot, EmausBot, ZéroBot, StringTheory11, Mentibot, Dsperlich, CeraBot, Galactic Messiah, DerekWinters, Fisherv, KasparBot and Anonymous: 42

- **Photon** *Source:* https://en.wikipedia.org/wiki/Photon?oldid=683740816 *Contributors:* AxelBoldt, WojPob, Mav, Bryan Derksen, The Anome, Tarquin, Koyaanis Qatsi, Ap, Josh Grosse, Ben-Zin~enwiki, Heron, Youandme, Spiff~enwiki, Bdesham, Michael Hardy, Ixfd64, TakuyaMurata, NuclearWinter, Looxix~enwiki, Snarfies, Ahoerstemeier, Stevenj, Julesd, Glenn, AugPi, Mxn, Smack, Pizza Puzzle, Wikiborg, Reddi, Lfh, Jitse Niesen, Kbk, Laussy, Bevo, Shizhao, Raul654, Jusjih, Donarreiskoffer, Robbot, Hankwang, Fredrik, Eman, Sanders muc, Altenmann, Bkalafut, Merovingian, Gnomon Kelemen, Hadal, Wereon, Anthony, Wjbeaty, Giftlite, Art Carlson, Herbee, Xerxes314, Everyking, Dratman, Michael Devore, Bensaccount, Foobar, Jaan513, DÅ‚ugosz, Zeimusu, LucasVB, Beland, Setokaiba, Kaldari, Vina, RetiredUser2, Icairns, Lumidek, Zondor, Randwicked, Eep², Chris Howard, Zowie, Naryathegreat, Discospinster, Rich Farmbrough, Yuval madar, Pjacobi, Vsmith, Ivan Bajlo, Dbachmann, Mani1, SpookyMulder, Kbh3rd, RJHall, Ben Webber, El C, Edwinstearns, Laurascudder, RoyBoy, Spoon!, Dalf, Drhex, Bobo192, Foobaz, I9Q79oL78KiL0QTFHgyc, La goutte de pluie, Zr40, Apostrophe, Minghong, Rport, Alansohn, Gary, Sade, Corwin8, PAR, UnHoly, Hu, Caesura, Wtmitchell, Bucephalus, Max rspct, BanyanTree, Cal 1234, Count Iblis, Egg, Dominic, Gene Nygaard, Ghirlandajo, Kazvorpal, UTSRelativity, Falcorian, Drag09, Boothy443, Richard Arthur Norton (1958-), Woohookitty, Linas, Gerd Breitenbach, StradivariusTV, Oliphaunt, Cleonis, Pol098, Ruud Koot, Mpatel, Nakos2208~enwiki, Dbl2010, Ch'marr, SDC, CharlesC, Alan Canon, Reddwarf2956, Mandarax, BD2412, Kbdank71, Zalasur, Sjakkalle, Rjwilmsi, Саша Стефановиħ, Strait, MarSch, Dennis Estenson II, Trlovejoy, Mike Peel, HappyCamper, Bubba73, Brighterorange, Cantorman, Egopaint, Noon, Godzatswing, FlaBot, RobertG, Arnero, Mathbot, Nihiltres, Fresheneesz, TeaDrinker, Srleffler, BradBeattie, Chobot, Jaraalbe, DVdm, Elfguy, EamonnPKeane, YurikBot, Bambaiah, Splintercellguy, Jimp, RussBot, Supasheep, JabberWok, Wavesmikey, KevinCuddeback, Stephenb, Gaius Cornelius, Salsb, Trovatore, Długosz, Tailpig, Joelr31, SCZenz, Randolf Richardson, Ravedave, Tony1, Roy Brumback, Gadget850, Dna-webmaster, Enormousdude, Lt-wiki-bot, Oysteinp, JoanneB, Ligart, John Broughton, GrinBot~enwiki, Sbyrnes321, Itub, SmackBot, Moeron, Incnis Mrsi, KnowledgeOfSelf, CelticJobber, Melchoir, Rokfaith, WilyD, Jagged 85, Jab843, Cessator, AnOddName, Skizzik, Dauto, JSpudeman, Robin Whittle, Ati3414, Persian Poet Gal, MK8, Jprg1966, Complexica, Sbharris, Colonies Chris, Ebertek, WordLife565, Vladis1av, RWincek, Aces lead, Stangbat, Cybercobra, Valenciano, EVula, A.R., Mini-Geek, AEM, DMacks, N Shar, Sadi Carnot, FlyHigh, The Fwanksta, Drunken Pirate, Yevgeny Kats, Lambiam, Harryboyles, IronGargoyle, Ben Moore, A. Parrot, Mr Stephen, Fbartolom, Dicklyon, SandyGeorgia, Mets501, Ceeded, Ambuj.Saxena, Ryulong, Vincecate, Astrobayes, Newone, J Di, Lifeveryatwhere, Tawkerbot2, JRSpriggs, Chetvorno, Luis A. Veguilla-Berdecia, CalebNoble, Xod, Gregory9, CmdrObot, Wafulz, Van helsing, John Riemann Soong, Rwflammang, Banedon, Wquester, Outriggr, Logical2u, Myasuda, Howardsr, Cydebot, Krauss, Kanags, A876, WillowW, Bvcrist, Hyperdeath, Hkyriazi, Rracecarr, Difluoroethene, Edgerck, Michael C Price, Tawkerbot4, Christian75, Ldussan, RelHistBuff, Waxigloo, Kozuch, Thijs!bot, Epbr123, Opabinia regalis, Markus Pössel, Mglg, 24fan24, Headbomb, Newton2, John254, J.christianson, Escarbot, Stannered, AntiVandalBot, Luna Santin, Jtrain4469, Normanmargolus, Tyco.skinner, TimVickers, NSH001, Dodecahedron~enwiki, Tim Shuba, Gdo01, Sluzzelin, Abyssoft, CosineKitty, AndyBloch, Bryanv, ScottStearns, Hroðulf, Bongwarrior, VoABot II, B&W Anime Fan, SHCarter, Lgoger, I JethroBT, Dirac66, Hveziris, Maliz, Lord GaleVII, TRWBW, Shijualex, Glen, DerHexer, Patstuart, Gwern, Taborgate, MartinBot, MNAdam, Jay Litman, HEL, Ralf 58, J.delanoy, DrKiernan, Trusilver, C. Trifle, AstroHurricane001, Numbo3, Pursey, CMDadabo, Kevin aylward, UchihaFury, Pirate452, H4xx0r, Iamthewalrus35, Iamthewalrus36, Gee Eff, Chimpy07, Dirkdiggler69, Lk69, Hallamfm, Annoying editter, Yehoodig, Acalamari, Foreigner1, McSly, Samtheboy, Tarotcards, Rominandreu, ARTE, Tanaats, Potatoswatter, Y2H, Divad89, Scott Illini, Stack27, THEblindwarrior, VolkovBot, AlnoktaBOT, Hyperlinker, DoorsAjar, TXiKiBoT, Oshwah, Cosmic Latte, The Original Wildbear, Davehi1, Chiefwaterfall, Vipinhari, Hqb, Anonymous Dissident, HansMair, Predator24, BotKung, Luuva, Calvin4986, Improve~enwiki, Kmhkmh, Richwil, Antixt, Gorank4, Falcon8765, GlassFET, Cryptophile, MattiasAndersson, AlleborgoBot, Carlodn6, NHRHS2010, Relilles~enwiki, Tpb, SieBot, Timb66, Graham Beards, WereSpielChequers, ToePeu.bot, JerrySteal, Android Mouse, Likebox, RadicalOne, Paolo.dL, Lightmouse, PbBot, Spartan-James, Duae Quartunciae, Hamiltondaniel, StewartMH, Dstebbins, ClueBot, Bobathon71, The Thing That Should Not Be, Mwengler, EoGuy, Jagun, RODERICKMOLASAR, Wwheaton, Dmlcyal8er, Razimantv, Mild Bill Hiccup, Feebas factor, J8079s, Rotational, MaxwellsLight, Awickert, Ex_cirial, PixelBot, Sun Creator, NuclearWarfare, PhySusie, El bot de la dieta, DerBorg, Shamanchill, PoofyPeter99, J1.grammar natz, Laserheinz, TimothyRias, XLinkBot, Jovianeye, Petedskier, Hess88, Addbot, Mathieu Perrin, DOI bot, DougsTech, Download, James thirteen, AndersBot, LinkFA-Bot, Barak Sh, AgadaUrbanit, Тивеρополник, Dayewalker, Quantumobserver, Kein Einstein, Legobot, Luckas-bot, Yobot, Kilom691, Allowgolf~enwiki, AnomieBOT, Ratul2000, Kingpin13, Materialscientist, Citation bot, Xqbot, Ambujarind69, Mananay, Emezei, Sharhalakis, Shirik, RibotBOT, Rickproser, SongRenKai, Max derner, Merrrr, A. di M., ▢▢, CES1596, Paine Ellsworth, Gsthae with tempo!, Nageh, TimonyCrickets, WurzelT, Steve Quinn, Spacekid99, Radeksonic, Citation bot 1, Pinethicket, I dream of horses, HRoestBot, Tanweer Morshed, Eno crux, Tom.Reding, Jschnur, RedBot, IVAN3MAN, Gamewizard71, FoxBot, TobeBot, Earthandmoon, PleaseStand, Marie Poise, RjwilmsiBot, Антон Гліністы, Ripchip Bot, Ofercomay, Chemyanda, EmausBot, Bookalign, WikitanvirBot, Roxbreak, Word2need, Gcastellanos, Tommy2010, Dcirovic, K6ka, Hhhippo, Cogiati, 1howadsr1, StringTheory11, Waperkins, Jojojlj, Access Denied, Quondum, AManWithNoPlan, Raynor42, L Kensington, Maschen, HCPotter, Haiti333, RockMagnetist, Rocketrod1960, ClueBot NG, JASMEET SINGH HAFIST, Schicagos, Snotbot, Vinícius Machado Vogt, Helpful Pixie Bot, SzMithrandir, Bibcode Bot, BG19bot, Roberticus, Paolo Lipparini, Wzrd1, Rifath119, Davidiad, Mark Arsten, Peter.sujak, Wikarchitect, Hamish59, Caypartisbot, Penguinstorm300, KSI ROX, Bhargavuk1997, Chromastone1998, TheJJJunk, Nimmo1859, EagerToddler39, Dexbot, EZas3pt14, Webclient101, Chrisanion, Vanquisher.UA, Tony Mach, PREMDASKANNAN, Meghas, Reatlas, Profb39, Zerberos, Thesuperseo, The User 111, Eyesnore, Ybidzian, Tentinator, Illusterati, Celso ad, Quenhitran, Manul, DrMattV, Anrnusna, Wyn.junior, K0RTD, Monkbot, Vieque, BethNaught, Markmizzi, Garfield Garfield, Smokey2022, Zargol Rejerfree, RAL2014, Shahriar Kabir Pavel, Sdjncskdjnfskje, Anshul1908, Professor Flornoy, Thatguytestw, Tetra quark, Harshit100, KasparBot, Chinta 01, Geek3, TheKingOfPhysics and Anonymous: 497

- **Gluon** *Source:* https://en.wikipedia.org/wiki/Gluon?oldid=681546689 *Contributors:* AxelBoldt, CYD, Bryan Derksen, Gdarin, TakuyaMurata, Card~enwiki, Looxix~enwiki, Ellywa, Ahoerstemeier, Med, Schneelocke, Phys, Phil Boswell, Donarreiskoffer, Fredrik, Merovingian, Hadal, Giftlite, Herbee, Xerxes314, Eequor, Darrien, Keith Edkins, RetiredUser2, Icairns, Mike Rosoft, AlexChurchill, HedgeHog, Kenny TM~~enwiki, David Schaich, Ioliver, Mashford, El C, Kwamikagami, Ardric47, Obradovic Goran, Alansohn, Guy Harris, Dachannien, Ricky81682, Batmanand, Velella, Kazvorpal, April Arcus, Forteblast, Mpatel, Palica, BD2412, Kbdank71, Rjwilmsi, Macumba, Strait, Mike Peel, Bubba73, Klortho, FlaBot, Srleffler, Chobot, YurikBot, Wavelength, Bambaiah, Hairy Dude, Jimp, JabberWok, Zelmerszoetrop, Salsb, SCZenz, Randolf Richardson, Ravedave, Danlaycock, Bota47, LeonardoRob0t, Anclation~enwiki, Physicsdavid, Erudy, GrinBot~enwiki, Kgf0, SmackBot, Melchoir, Cessator, Benjaminevans82, Abtal, MK8, Colonies Chris, Can't sleep, clown will eat me, Decltype, Qcdmaestro, Edconrad, Darkpoison99, FredrickS, Omsharan, Pegasusbot, Gregbard, ProfessorPaul, Thijs!bot, Headbomb, Rriegs, Oreo Priest, AntiVandalBot, Shambolic Entity, Deflective, Mujokan, Yill577, Happycool, Mother.earth, Martynas Patasius, WiiWillieWiki, HEL, Hans Dunkelberg, Gombang, Inwind, Sheliak, Jonthaler, VolkovBot, TXiKiBoT, Davehi1, Kriak, Anonymous Dissident, Imasleepviking, AlleborgoBot, EJF, SieBot, Steven Crossin, OKBot, ClueBot, Wwheaton, Qsaw, Nucularphysicist, Ottava Rima, Gordon Ecker, Rhododendrites, Brews ohare, Cacadril, RexxS, JKeck, Against the current, SkyLined, Addbot, DOI bot, Lightbot, Skippy le Grand Gourou, Luckas-bot, Planlips, AnomieBOT, Jim1138, JackieBot, Citation bot Bci2, ArthurBot, Xqbot, Neil95, Triclops200, Omnipaedista, TorKr, ▢▢, Paine Ellsworth, Ivoras, Citation bot 1, Pekayer11, Rameshngbot, PNG, RjwilmsiBot, TjBot, Lilcal89012, EmausBot, Socob, JSquish, StringTheory11, Quondum, TyA, Maschen, RolteVolte, ClueBot NG, Timothy jordan, Maplelanefarm, Bibcode Bot, BG19bot, Gravitoweak, Cadiomals, Tropcho, Fraulein451, DrHjmHam, Rhlozier, D.shinkaruk, Yaara dildaara, BronzeRatio, Monkbot, Yikkayaya, KasparBot and Anonymous: 142

- **W and Z bosons** *Source:* https://en.wikipedia.org/wiki/W_and_Z_bosons?oldid=676803444 *Contributors:* AxelBoldt, Sodium, Mav, Bryan Derksen, The Anome, Ap, Andre Engels, Danny, Roadrunner, DrBob, Michael Hardy, Tim Starling, Karada, Egil, Ahoerstemeier, Ryan Cable, Julesd, Mxn, Charles Matthews, Ike9898, Saltine, Phys, Topbanana, BenRG, Finlay McWalter, Twang, Phil Boswell, Donarreiskoffer, Robbot, Pigsonthewing, Nurg, DHN, Xanzzibar, M-Falcon, Giftlite, Tremolo, Harp, Herbee, Xerxes314, Jeremy Henty, Bodhitha, LiDaobing, RetiredUser2, Icairns, Mike Rosoft, Vsmith, Gianluigi, Kjoonlee, Drhex, Obradovic Goran, Jérôme, Fkbreitl, Cameron.simpson, Gene Nygaard, Linas, LoopZilla, Graham87, Kbdank71, Rjwilmsi, Strait, Mike Peel, Lmatt, Goudzovski, Chobot, FrankTobia, Roboto de Ajvol, Ugha, Mushin, Bambaiah, Wester, Hairy Dude, Hellbus, Salsb, Seb35, Długosz, Turbolinux999, Ravedave, Scottfisher, Dna-webmaster, Modify, Argo Navis, Teply, Sbyrnes321, SmackBot, Tom Lougheed, Jagged 85, ZerodEgo, Dauto, Bluebot, Shaggorama, Sbharris, Niels Olson, Radagast83, Acdx, John, Lottamiata, Happy-melon, Tubezone, MightyWarrior, Joelholdsworth, Tangobot, Michael C Price, Quibik, Dchristle, Realjanuary, Headbomb, Davidhorman, Nosirrom, Certain, Gökhan, JAnDbot, Tigga, Omeganian, Brimofinsanity, TheEditrix2, Trapezoidal, Magioladitis, ThoHug, Leyo, Lilac Soul, HEL, Rod57, Y2H, HiEv, Adam Zivner, Madblueplanet, Sheliak, Dextrose, Anonymous Dissident, Synthebot, Antixt, Coronellian~enwiki, SieBot, STANMAR725, Jim E. Black, Gerakibot, Martin Kealey, CutOffTies, Fratrep, ClueBot, Mild Bill Hiccup, Alexbot, Carsrac, SkyLined, Dieppu, Stephen Poppitt, Addbot, Eric Drexler, Toyokuni3, Mjamja, Ronkonkaman, Download, CarsracBot, ChenzwBot, Lightbot, M sotirov, Luckas-bot, Yobot, Jim1138, MehrdadAfshari, ArthurBot, Ernsts, A. di M., Howard McCay, FrescoBot, Paine Ellsworth, D'ohBot, Citation bot 1, Gil987, Tom.Reding, Swallerick, FoxBot, Earthandmoon, Tm1729, TjBot, Антон Гліністы, Newty23125, EmausBot, Mnkyman, StringTheory11, Quondum, MisterDub, WaterCrane, Whoop whoop pull up, ClueBot NG, Helpful Pixie Bot, Bibcode Bot, BG19bot, Bakkedal, JYBot, Mamaphyskerin, Anrnusna, MartinNicklin, Boidal-Quantized and Anonymous: 137

- **Scalar boson** *Source:* https://en.wikipedia.org/wiki/Scalar_boson?oldid=653639224 *Contributors:* AugPi, Phys, Giftlite, Linas, Mpatel, RussBot, Bhny, SmackBot, Sergio.ballestrero, QFT, WLevine, Magioladitis, Tdadamemd, Sigmundur, Antixt, Jim E. Black, JL-Bot, ClueBot, Naradawickramage, Addbot, Dr. Universe, Anypodetos, J04n, Erik9bot, Fortdj33, Tomville219, RedBot, GoingBatty, Carbosi, ZéroBot, RolteVolte, Darine Of Manor, Parcly Taxel, Beaumont877, Mogism and Anonymous: 15

- **Higgs boson** *Source:* https://en.wikipedia.org/wiki/Higgs_boson?oldid=683542412 *Contributors:* AxelBoldt, CYD, ClaudeMuncey, Bryan Derksen, Manning Bartlett, Roadrunner, David spector, Heron, Ewen, Stevertigo, Edward, Boud, TeunSpaans, Dante Alighieri, Ixfd64, Gau-

BG19bot, Scottaleger, Mcarmier, Jibu8, Loupatriz67, Dave4478, Frze, Ervin Goldfain, Reader505, Mark Arsten, Lovetrivedi, BarbaraMervin, Silvrous, Drcooljoe, Cadiomals, Joydeep, Altaïr, Piet De Pauw, Jeancey, Sovereign8, Visuall, Ownedroad9, Brainssturm, Jw2036, Writ Keeper, DPL bot, Nickni28, Philpill691, Lee.boston, Scientist999, Benjiboy187, Duxwing, Cengime, Skiret girdet njozet, GRighta, Downtownclaytonbrown, Diasjordan, Ghsetht, Marioedesouza, BattyBot, 1narendran, LORDCOTTINGHAM2, NO SOPA, Tchaliburton, Wijnburger, StarryGrandma, Mdann52, Dilaton, Magikal Samson, Samuelled, Dja1979, Georgegroom, BecurSansnow, EuroCarGT, MSUGRA, Rhlozier, Pscott558, Turullulla, Blueprinteditor, Misterharris~enwiki, AstroDoc, Bigbear213, Dexbot, Randomizer3, Daggerbot, DoctorLazarusLong, Caroline1981, Nitpicking polisher, SoledadKabocha, Gsmanu007, Windows.dll, Mogism, Prabal123koirala, Abitoby, Clidog, Rongended, Darryl from Mars, Cerabot~enwiki, MuonRay, TheTruth72, Capt. Mohan Kuruvilla, Gatheringstorm2, Jason7898, Mumbai999999, SkepticalKid, Cjean42, Nmrzuk, Lugia2453, Mafuee, Frosty, SFK2, Thegodparticlebook, Rijensky, Mishra866868, Rockstar999999999, Toddbeck911, Nilaykumar07, Thepalerider2012, WikiPhysTech, The Anonymouse, Ahmar Saeed, Pincrete, Apidium23, Prahas.wiki, Exenola, Pdotpwns, Epicgenius, Fireballninja, Greengreengreenred, 󠀀󠀀, Technogeek101, NicoPosner, Apurva Godghase, Durfyy, Soumya Mittal, American In Brazil, SaifAli13, Qwerkysteve, Spatiandas, Retroherb, Tango303, Hoppeduppeanut, Redplain, Shaelote, Quadrum, AntiguanAcademic, Simpsonojsim, Agyeyaankur, DavidLeighEllis, Ethanthevelociraptor, Qfang12, Comp.arch, Eletro1903, E8xE8, HeineBOB, Kahtar, Depthdiver, JAaron95, Mfb, Anrnusna, Stamptrader, Man of Steel 85, Cteirmn, AiraCobra, MyNameIsn'tElvis, Meganlock8, Sxxximf, Drsoumyadeepb, 22merlin, Ndidi Okonkwo Nwuneli, Monkbot, Dialga5555, Fred1810, Akro7, Pewpewpewpapapa, BradNorton1979, 21bhargav, Whistlemethis, Thinking Skeptically, Amk365, Gagnonlg, Knowledgebattle, L21234, TheNextMessiah, Naterealm224, Joey van Helsing, Adrian Lamplighter, Arnab santra, Gemadi, BATMAN1021, Isambard Kingdom, Mercedes321, DrKitts, KasparBot, JJMC89, GBjun3, TheRoamer64, Firstcause, Seventhorbitday, RobeDM, Wordfunk and Anonymous: 963

- **Graviton** *Source:* https://en.wikipedia.org/wiki/Graviton?oldid=677652863 *Contributors:* CYD, Bryan Derksen, Timo Honkasalo, XJaM, Fubar Obfusco, Maury Markowitz, Kaczor~enwiki, Jketola, TakuyaMurata, Eric119, Looxix~enwiki, Glenn, Cyan, Wooster, Charles Matthews, Timwi, Wik, BenRG, Donarreiskoffer, Scott McNay, Stephan Schulz, Arkuat, Chris Roy, Merovingian, Davidl9999, Giftlite, Xerxes314, Jason Quinn, Matt Crypto, CryptoDerk, RetiredUser2, Icairns, Zfr, Lumidek, Ukexpat, Urvabara, Discospinster, Pjacobi, Vapour, Brian0918, El C, Joanjoc~enwiki, Dalf, Army1987, Mpvdm, La goutte de pluie, Physicistjedi, Daniel Arteaga~enwiki, Zenosparadox, Dethtron5000, Keenan Pepper, Viridian, SidP, Falcorian, Skeejay, Simetrical, Dr Archeville, Mpatel, Kyleca, Tmassey, Christopher Thomas, Tevatron~enwiki, Kbdank71, Nightscream, Koavf, Mike Peel, Ems57fcva, FlaBot, RexNL, Chobot, DVdm, Roboto de Ajvol, Spacepotato, Anonymous editor, SnoopY~enwiki, Salsb, Bachrach44, Hyperbrand, NickBush24, Pnrj, RL0919, EEMIV, IslandGyrl, Bota47, C h fleming, Petri Krohn, Mario23, Alias Flood, Tim314, Teply, GrinBot~enwiki, SmackBot, Amcbride, Melchoir, Eskimbot, Gilliam, Skizzik, Timneu22, Complexica, Villarinho, Colonies Chris, V1adis1av, Chlewbot, Xyzzyplugh, Jmnbatista, Fuhghettaboutit, Sadi Carnot, Yevgeny Kats, TenPoundHammer, Lambiam, Zaphraud, JorisvS, Mr Stephen, Ramuman, Quasar Jarosz, Lottamiata, Firewall62, Kurtan~enwiki, CmdrObot, BeenAroundAWhile, WeggeBot, Shultz IV, UncleBubba, Michael C Price, Anthmoo, Thijs!bot, Epbr123, Headbomb, KevinS06, Opelio, Spartaz, JAnDbot, Xoneca, SHCarter, Pikazilla, Robin S, STBot, Kostisl, J.delanoy, Tarotcards, Coppertwig, Wesino, Sava ankit2006, Tygrrr, Idioma-bot, Sheliak, JoAnneThrax, TXiKiBoT, WilliamSommerwerck, Hqb, Anonymous Dissident, Antixt, SieBot, Flyer22, Henry Delforn (old), ClueBot, Ergn, Darkicebot, DenverRedhead, Addbot, Eric Drexler, Uruk2008, DOI bot, BrianBop, PJonDevelopment, F Notebook, Legobot, Picturesofnothing, Dov Henis, Alfredschrader, Eric-Wester, AnomieBOT, VanishedUser sdu9aya9fasdsopa, Jim1138, Materialscientist, Citation bot, Tomflaherty, ProtectionTaggingBot, Waleswatcher, FrescoBot, Juto20, LucienBOT, Paine Ellsworth, I dream of horses, Tom.Reding, RedBot, Omar.tigereyes, IVAN3MAN, Ashish.kotwal, Michael9422, D0wnfalle, EmausBot, Octaazacubane, 8digits, Slightsmile, K6ka, Thecheesykid, User10 5, Rcsprinter123, Orbjeeples, Puffin, Herk1955, ClueBot NG, Raidr, Masssly, Helpful Pixie Bot, Bibcode Bot, BG19bot, Shapoopy178, ServiceAT, PhnomPencil, Trevayne08, Brainssturm, Tjamcclain2, ChrisGualtieri, Ariscod, TheUyulala, LightandDark2000, Jessybun, Makecatbot, Kryomaxim, JRYon, Andyhowlett, Mark viking, Yorsh07, CensoredScribe, WPratiwi, Monkbot, Bryan Paul Senior, Dr.Begich, Nompynuthead, Jacobflarsen and Anonymous: 196

- **Bose–Einstein condensate** *Source:* https://en.wikipedia.org/wiki/Bose%E2%80%93Einstein_condensate?oldid=683698129 *Contributors:* Kpjas, CYD, Archibald Fitzchesterfield, Bryan Derksen, Olof, Tarquin, Gareth Owen, Josh Grosse, Hfastedge, Spiff~enwiki, Michael Hardy, Gabbe, TakuyaMurata, SebastianHelm, Alfio, Ellywa, Cyp, Stevan White, Darkwind, Glenn, Mxn, Schneelocke, Loren Rosen, Feedmecereal, Dino, Wikiborg, The Anomebot, ElusiveByte, BenRG, JorgeGG, Donarreiskoffer, Chris 73, Nurg, Robinh, GreatWhiteNortherner, Dave6, MFalcon, Matt Gies, Giftlite, Smjg, Inter, Herbee, Dratman, Tom-, Eequor, Balenman, Chrissmith, Mooquackwooftweetmeow, Toytoy, XxPantherNovaXx, Fangz, Piotrus, Karol Langner, Brian Jackson, Spiralhighway, Sam Hocevar, Kramer, Nickptar, Vivacissamamente, Grunt, Eep², NightMonkey, Lone Isle, Noisy, Discospinster, Guanabot, ThomasK, Vsmith, Aardark, Paul August, Bender235, TOR, RJHall, El C, Lycurgus, Ruyn, Laurascudder, Jpgordon, Fuxx, Directorstratton, Slicky, Sasquatch, Haham hanuka, Alansohn, Arthena, Nwinther, PAR, Pion, Kfitzgib, Cjnm, Tom12519, Snowolf, Einstein9073, BRW, KapilTagore, Pauli133, Gene Nygaard, Joriki, OwenX, Woohookitty, Linas, David Haslam, Benbest, Ruud Koot, Jeff3000, Astrophil, BlaiseFEgan, Bugman, Sjö, Rjwilmsi, Amire80, Rillian, BlueMoonlet, Salix alba, Keimzelle, Exeunt, Azure8472, FlaBot, SchuminWeb, The.valiant.paladin, Shade², Pete.Hurd, Srleffler, Erik4, King of Hearts, Chobot, DVdm, Sasoriza, YurikBot, Wavelength, Taurrandir, Rob T Firefly, Hairy Dude, Huw Powell, Flameviper, Michael Slone, JabberWok, David Woodward, Shell Kinney, WulfTheSaxon, Truetyper, Howcheng, Chakazul, Katrielalex, Dogcow, Grafikm fr, Zwobot, Wangi, Wknight94, FF2010, 2over0, Closedmouth, Dr.alf, Stuhacking, Otto ter Haar, Groyolo, Allium, SmackBot, Serg3d2, RossyMiles, Oxford Comma, Olegt1, Nickst, Eskimbot, Dilbert3, Gaff, Gilliam, Kmarinas86, Thumperward, Hichris, DHN-bot~enwiki, Raistuumum, Salmar, Tcb Beany, Karpita, Can't sleep, clown will eat me, Nick Levine, Kelvin Case, Neo139, Onorem, MBlume, Voyajer, Rrburke, Xyzzyplugh, GeorgeMoney, TedE, Bigmantonyd, Rich.lewis, DMacks, Xiutwel, Ligulembot, Mion, Sadi Carnot, Josellis, Tethros, Lucretius~enwiki, Lambiam, Andi47, John, MagnaMopus, Jaganath, SteveG23, Mgiganteus1, Ckatz, BillFlis, Kyoko, Dicklyon, Inquisitus, Phuzion, Brienanni, Iridescent, JMK, Clarityfiend, FelisSchrödingeris, Frank Lofaro Jr., CRGreathouse, ZICO, BeenAroundAWhile, DSachan, Orannis, Myasuda, Leakeyjee, Equendil, Stebbins, Kanags, MC10, Tashafairbairn, Mato, Gogo Dodo, JFreeman, Mattjball, Omicronpersei8, Thijs!bot, Epbr123, Wikid77, Trevheg, Fiction Alchemist, Sam Van Kooten, Headbomb, Second Quantization, Iviney, CharlotteWebb, AntiVandalBot, 17Drew, Gökhan, MSBOT, Boleslaw, Sinnerwiki, Sophosmoros, Magioladitis, WolfmanSF, VoABot II, Sushant gupta, Bakken, Ggorelik, Tonyfaull, BatteryIncluded, Dirac66, David Eppstein, LorenzoB, Talon Artaine, Torsionalmetric, Starryharlequin, N734LQ, Anonymous 57, Sketchjoy, Custos0, J.delanoy, MITBeaverRocks, Jtw11, Bogey97, Maurice Carbonaro, AquamarineOnion, Glaux, AppleMacReporter, AntiSpamBot, Tendays, Enix150, Neil Dodgson, Idioma-bot, Austinmohr, Gnipahellir, A4bot, Qxz, Martin451, Mitchell26, Natural Philosopher, Mazarin07, Akhuettel, Spinningspark, Kapalama, Cryonic07, PaddyLeahy, Biscuittin, Awemond, WereSpielChequers, Cmossol, Matthew Yeager, Deathgleaner, Reuqr, Likebox, JD554, Reinderien, Topher385, Scorpion451, Lightmouse, Jakeng, Coldcreation, Psycherevolt, Melcombe, AllHailZeppelin, Crazz bug 5,

Martarius, ClueBot, MonkeyMensch, Snigbrook, The Thing That Should Not Be, EoGuy, Emil70, Zero over zero, Razimantv, Maymay, The-generalguy, DrakeUnlimited, NuclearWarfare, Dboiko, Doktor Mephisto, SchreiberBike, Thingg, Jonverve, SoxBot III, Egmontaz, DumZiBoT, Ost316, Rreagan007, SilvonenBot, SkyLined, MaizeAndBlue86, Csingh23592, Addbot, DOI bot, Jojhutton, Miskaton, Friginator, Download, ChenzwBot, AtheWeatherman, 84user, Tide rolls, Lightbot, OlEnglish, Teles, SPat, Megaman en m, Ben Ben, Yobot, Wireader, VectorField, AnomieBOT, TheUfoFiles, Aaagmnr, Materialscientist, Limideen, Citation bot, MetaplecticGroup, Natural RX, Xqbot, BME-physics, Lunain-tern, RibotBOT, Verbum Veritas, Nixón, HJ Mitchell, Quantum 235, Citation bot 1, Maan361, Gil987, Gaba p, I dream of horses, Coekon, RedBot, Akalabeth, Keri, Asrrin29, Senra, Canuckian89, JV Smithy, DARTH SIDIOUS 2, Obankston, Hajatvrc, Nkf31, EmausBot, John of Reading, Gfoley4, Physics16, GoingBatty, KHamsun, Solarra, Lent1999, H3llBot, Quondum, Timetraveler3.14, DougEFresh1122, Donner60, Fairskys, Carmichael, Jalexander-WMF, ClueBot NG, Gareth Griffith-Jones, Movses-bot, All Hail Hypnotoad!, Zak.estrada, Widr, Fqr2010, Helpful Pixie Bot, HMSSolent, Jubobroff, Bibcode Bot, BG19bot, Northamerica1000, AvocatoBot, Wowwii, Rm1271, Mr.viktor.stepanov, BattyBot, Mrt3366, ChrisGualtieri, Adwaele, Baileybrooks, FlappyJenkins, Dexbot, Makecat-bot, Baldoc83, Jamesx12345, Stewwie, Avra-hamleib, Sakurai23, Marcela louis, Reatlas, Epicgenius, Nonsenseferret, Aarya19991111, Ginsuloft, Aritcle, KillerKira, John Doppler, Aarjun Rampal, Happy Attack Dog, Arnaud Migres, Tusharkashyap2001, AfrikanischePost, Hans8654, Sumandark8600, Shengxingwu, Antsiepantsie, Yohoona, Mysterious Gopher, KasparBot, Liamsmith12 and Anonymous: 500

- **Bosonic field** *Source:* https://en.wikipedia.org/wiki/Bosonic_field?oldid=663937028 *Contributors:* Michael Hardy, Zocky, Giftlite, Fropuff, Danski14, Ittiz, SmackBot, Colonies Chris, QFT, JorisvS, Headbomb, Jed1978, Venny85, Jamessmithpage, ClueBot, Nilradical, Jovianeye, TStein, Kenneth Dawson, Helpful Pixie Bot, MarkovianStumble and Anonymous: 3

- **List of particles** *Source:* https://en.wikipedia.org/wiki/List_of_particles?oldid=682746251 *Contributors:* AxelBoldt, Danny, Rmhermen, Stevertigo, Bdesham, Ahoerstemeier, Stan Shebs, Docu, Salsa Shark, Nikai, Evercat, Schneelocke, Charles Matthews, Jitse Niesen, CBDunker-son, Bevo, Raul654, Donarreiskoffer, Robbot, Sanders muc, Merovingian, Pengo, Giftlite, Herbee, Xerxes314, Dratman, Jeremy Henty, Alen-sha, Bodhitha, Physicist, Hayne, Quadell, RetiredUser2, Mysidia, Icairns, Asbestos, D6, Urvabara, Discospinster, Rich Farmbrough, FT2, Qutezuce, ArnoldReinhold, Neko-chan, El C, Laurascudder, Susvolans, EmilJ, Physicistjedi, Minghong, Gbrandt, Eddideigel, Axl, Mac Davis, David Ko, Radical Mallard, RJFJR, Count Iblis, Dirac1933, TenOfAllTrades, LFaraone, Oleg Alexandrov, Linas, JarlaxleArtemis, Dun-can.france, GregorB, Cedrus-Libani, Karam.Anthony.K, Palica, Rjwilmsi, Zbxgscqf, JLM~enwiki, Strait, Ems57fcva, Krash, Dan Guan, Dan-nyWilde, Lmatt, Goudzovski, Chobot, YurikBot, Bambaiah, Vuvar1, Madkayaker, Hydrargyrum, Presscorr, Chaos, Salsb, Tavilis, SCZenz, Lexicon, TUSHANT JHA, Dna-webmaster, Tomvds, Poulpy, Cstmoore, TLSuda, NeilN, MacsBug, Tom Lougheed, McGeddon, Bazza 7, WookieInHeat, Derdeib, Yamaguchi⬚⬚, Betacommand, Bluebot, Master of Puppets, DHN-bot~enwiki, Raistuumum, Juancnuno, Kittybrew-ster, Acepectif, Ligulembot, TriTertButoxy, ArglebargleIV, Khazar, John, FrozenMan, JorisvS, 041744, Dr Greg, Slakr, Mets501, Scor-pion0422, Cbuckley, Iridescent, TwistOfCain, Happy-melon, JRSpriggs, Flickboy, Van helsing, Lithium6, Neelix, Rotiro, Cydebot, Quibik, Christian75, Omicronpersei8, Thijs!bot, Qwyrxian, TauLibrus, Headbomb, Inner Earth, 49, Guptasuneet, Scottmsg, WinBot, Elmoosecapi-tan, Tyco.skinner, AubreyEllenShomo, Arch dude, Johnman239, Mwarren us, TheEditrix2, CalamusFortis, MartinBot, Sadisticsuburbanite, Bissinger, Anaxial, CommonsDelinker, Maurice Carbonaro, Zojj, OliverHarris, Joshmt, Adanadhel, Lseixas, Graphite Elbow, VolkovBot, Jmrowland, Quilbert, Anonymous Dissident, Dstary, Escalona, JPMasseo, Figureskatingfan, Inx272, Meters, Antixt, Hamish a e fowler, God-dersUK, Bluetryst, SieBot, Ishvara7, WereSpielChequers, Audrius u, VovanA, Paolo.dL, RSStockdale, Anchor Link Bot, StewartMH, Ex-plicit, ClueBot, Unbuttered Parsnip, Nolimitownass, DragonBot, Atomic7732, TimothyRias, SkyLined, Addbot, DOI bot, Jojhutton, Favonian, LinkFA-Bot, OlEnglish, Teles, Legobot, Luckas-bot, Yobot, Dov Henis, Azcolvin429, AnomieBOT, Götz, Icalanise, Flewis, Materialscientist, OllieFury, Vuerqex, ArthurBot, Vulcan Hephaestus, Blennow, Reality006, Coretheapple, Jcimorra, RibotBOT, Ernsts, A. di M., Axelfoley12, Zosterops, FrescoBot, Paine Ellsworth, Citation bot 1, JIK1975, Tom.Reding, Diffequa, WikitanvirBot, Racerx11, 112358sam, Aegnor.erar, Hops Splurt, HESUPERMAN, Hhhippo, AvicBot, JSquish, StringTheory11, Waperkins, Bamyers99, Suslindisambiguator, L Kensington, Den-nisIsMe, RockMagnetist, ClueBot NG, Snotbot, Primergrey, Vio45lin, Widr, MsFionnuala, Oklahoma3477, Bibcode Bot, CityOfSilver, Cap'n G, BML0309, Dan653, Twocount, Penguinstorm300, Dexbot, LightandDark2000, Ohiggy, TwoTwoHello, Andyhowlett, Printersmoke, Orion 2013, ARUNEEK, Seino van Breugel, AspaasBekkelund, TheMagikCow, Vyom27, ParkersComments, Selva Ganapathy and Anonymous: 290

11.8.2 Images

- **File:2-photon_Higgs_decay.svg** *Source:* https://upload.wikimedia.org/wikipedia/commons/3/32/2-photon_Higgs_decay.svg *License:* CC BY-SA 3.0 *Contributors:* Own work *Original artist:* Parcly Taxel

- **File:4-lepton_Higgs_decay.svg** *Source:* https://upload.wikimedia.org/wikipedia/commons/b/b2/4-lepton_Higgs_decay.svg *License:* CC BY-SA 3.0 *Contributors:* Own work *Original artist:* Parcly Taxel

- **File:AIP-Sakurai-best.JPG** *Source:* https://upload.wikimedia.org/wikipedia/commons/2/2b/AIP-Sakurai-best.JPG *License:* Public domain *Contributors:* Own work *Original artist:* self

- **File:Ambox_important.svg** *Source:* https://upload.wikimedia.org/wikipedia/commons/b/b4/Ambox_important.svg *License:* Public domain *Contributors:* Own work, based off of Image:Ambox scales.svg *Original artist:* Dsmurat (talk · contribs)

- **File:Baryon_decuplet.svg** *Source:* https://upload.wikimedia.org/wikipedia/commons/f/f6/Baryon_decuplet.svg *License:* Public domain *Contributors:* Own work (Original text: *self-made*) *Original artist:* Wierdw123 at English Wikipedia

- **File:Beta_Negative_Decay.svg** *Source:* https://upload.wikimedia.org/wikipedia/commons/8/89/Beta_Negative_Decay.svg *License:* Public domain *Contributors:* This vector image was created with Inkscape. *Original artist:* Joel Holdsworth (Joelholdsworth)

- **File:Bohr-atom-PAR.svg** *Source:* https://upload.wikimedia.org/wikipedia/commons/5/55/Bohr-atom-PAR.svg *License:* CC-BY-SA-3.0 *Contributors:* Transferred from en.wikipedia to Commons. *Original artist:* Original uplo:JabberWok]] at en.wikipedia

- **File:Bose-Einstein_Condensation.ogv** *Source:* https://upload.wikimedia.org/wikipedia/commons/d/d9/Bose-Einstein_Condensation.ogv *License:* CC BY-SA 3.0 *Contributors:* Own work *Original artist:* Jubobroff Jubobroff J.Bobroff and full list in credits

- **File:Bose_Einstein_condensate.png** *Source:* https://upload.wikimedia.org/wikipedia/commons/a/af/Bose_Einstein_condensate.png *License:* Public domain *Contributors:* NIST Image *Original artist:* NIST/JILA/CU-Boulder

- **File:Question_book-new.svg** *Source:* https://upload.wikimedia.org/wikipedia/en/9/99/Question_book-new.svg *License:* Cc-by-sa-3.0 *Contributors:*
  Created from scratch in Adobe Illustrator. Based on Image:Question book.png created by User:Equazcion *Original artist:*
  Tkgd2007

- **File:SatyenBose1925.jpg** *Source:* https://upload.wikimedia.org/wikipedia/commons/f/fe/SatyenBose1925.jpg *License:* Public domain *Contributors:* Picture in Siliconeer *Original artist:* Unknown

- **File:Spontaneous_symmetry_breaking_(explanatory_diagram).png***Source:* https://upload.wikimedia.org/wikipedia/commons/a/a5/_symmetry_breaking_%28explanatory_diagram%29.png *License:* CC BY-SA 3.0 *Contributors:* Own work *Original artist:* FT2

- **File:Standard_Model_of_Elementary_Particles.svg** *Source:* https://upload.wikimedia.org/wikipedia/commons/0/00/Standard_Model_of_Elementary_Particles.svg *License:* CC BY 3.0 *Contributors:* Own work by uploader, PBS NOVA [1], Fermilab, Office of Science, United States Department of Energy, Particle Data Group *Original artist:* MissMJ

- **File:Stimulatedemission.png** *Source:* https://upload.wikimedia.org/wikipedia/commons/8/8a/Stimulatedemission.png *License:* CC-BY-SA-3.0 *Contributors:* en:Image:Stimulatedemission.png *Original artist:* User:(Automated conversion),User:DrBob

- **File:Stylised_Lithium_Atom.svg** *Source:* https://upload.wikimedia.org/wikipedia/commons/e/e1/Stylised_Lithium_Atom.svg *License:* CC-BY-SA-3.0 *Contributors:* based off of Image:Stylised Lithium Atom.png by Halfdan. *Original artist:* SVG by Indolences. Recoloring and ironing out some glitches done by Rainer Klute.

- **File:Symmetricwave2.png** *Source:* https://upload.wikimedia.org/wikipedia/commons/1/1d/Symmetricwave2.png *License:* CC BY 3.0 *Contributors:* Own work *Original artist:* TimothyRias

- **File:Vertex_correction.svg** *Source:* https://upload.wikimedia.org/wikipedia/commons/8/87/Vertex_correction.svg *License:* Public domain *Contributors:* ? *Original artist:* User:Harmaa

- **File:VisibleEmrWavelengths.svg** *Source:* https://upload.wikimedia.org/wikipedia/commons/e/e2/VisibleEmrWavelengths.svg *License:* Public domain *Contributors:* created by me *Original artist:* maxhurtz

- **File:Wikinews-logo.svg** *Source:* https://upload.wikimedia.org/wikipedia/commons/2/24/Wikinews-logo.svg *License:* CC BY-SA 3.0 *Contributors:* This is a cropped version of Image:Wikinews-logo-en.png. *Original artist:* Vectorized by Simon 01:05, 2 August 2006 (UTC) Updated by Time3000 17 April 2007 to use official Wikinews colours and appear correctly on dark backgrounds. Originally uploaded by Simon.

- **File:Wikiquote-logo.svg** *Source:* https://upload.wikimedia.org/wikipedia/commons/f/fa/Wikiquote-logo.svg *License:* Public domain *Contributors:* ? *Original artist:* ?

- **File:Wikisource-logo.svg** *Source:* https://upload.wikimedia.org/wikipedia/commons/4/4c/Wikisource-logo.svg *License:* CC BY-SA 3.0 *Contributors:* Rei-artur *Original artist:* Nicholas Moreau

- **File:Wiktionary-logo-en.svg** *Source:* https://upload.wikimedia.org/wikipedia/commons/f/f8/Wiktionary-logo-en.svg *License:* Public domain *Contributors:* Vector version of Image:Wiktionary-logo-en.png. *Original artist:* Vectorized by Fvasconcellos (talk · contribs), based on original logo tossed together by Brion Vibber

- **File:Young_Diffraction.png** *Source:* https://upload.wikimedia.org/wikipedia/commons/8/8a/Young_Diffraction.png *License:* Public domain *Contributors:* ? *Original artist:* ?

11.8.3 Content license

- Creative Commons Attribution-Share Alike 3.0

www.ingramcontent.com/pod-product-compliance
Lightning Source LLC
Chambersburg PA
CBHW080818180526
45168CB00006B/2497